CENTRIFUGATION TECHNIQUES

Centrifugation Techniques

SHALINEE NAIDOO

ARCLER PRESS

www.arclerpress.com

Centrifugation Techniques

SHALINEE NAIDOO

Arcler Press
2010 Winston Park Drive,
2nd Floor
Oakville, ON L6H 5R7
Canada
www.arclerpress.com
Tel: 001-289-291-7705

001-905-616-2116

Fax: 001-289-291-7601

Email: orders@arclereducation.com

ISBN: 978-1-77361-061-0 (Hardcover)

About the Author

Shalinee Naidoo obtained her Master of Science degree cum laude in Microbiology from The University of KwaZulu-Natal, Durban, South Africa in 2014. Her research interests lie in Environmental Microbiology and water management. She has been involved in a range of research projects funded by The Water Research Commission and The Medical Research Council of South Africa. She is currently a member of the Golden Key International Honour Society and a recipient of numerous scholarship programs including the Erasmus Mundus SAPIENT program between Europe and South Africa. She has spent time teaching Chemistry and English to children in Asia and is currently the founder of Scientist's Sanctuary, an online science communication company specialising in science writing and designing.

Contents

List of Figures

List of Tables

Preface

Cell biology involves the analysis of cells and tissues at both the morphological and biochemical levels in an attempt to understand the molecular mechanisms operating in any living organism. Proper biochemical analysis of biological samples is imperative for constructing complete and true cell biology. The subcellular organization of cells can be studied via electron microscopy or by the separation and subsequent molecular and biochemical analysis of individual cellular components. While electron microscopy provides detailed information about the appearance of cellular organelles, not much information can be obtained regarding their individual functions. As such, a way to separate the different organelles from remaining cellular components was needed (OpenLearn, 2017).

This was made possible by a technique known as cell fractionation which allows for the isolation of various cellular components (nuclear, cytosolic and mitochondrial) based on their subcellular compartmentalized distribution followed by analysis of their molecular information ranging from the state and distribution of a cell or subcellular component to their response to a range of treatments (Dimauro et al, 2012). The separation of cellular components and particles by sedimentation is one of the most powerful tools in biology. Even though sedimentation using centrifugation is not a new technology, it is essential for conducting cutting edge genomic and proteomic research, as it aids in providing purified particles of interest. As such, centrifugation forms one of the most important and widely applied research techniques used in various disciplines today. It services a range of research fields such as biochemistry, cellular and molecular biology while allowing large scale applications across a broad spectrum of industries ranging from large scale commercial applications in the pharmaceutical and agricultural industries to small scale laboratory-scale scientific studies.

Miescher was one of the first researchers to use a centrifuge for the isolation of a cell organelle when he separated nuclei from human pus cells in 1871. A survey conducted at the US National Institutes of Health revealed that over 65% of research workers believe that using centrifugation to purify cells, subcellular organelles, viruses, proteins, and nucleic acids is an essential part of their everyday research, further proving the importance of this technique in today's scientific world (Sigma-Aldrich, 2011). Current research applications rely on centrifugation for the isolation and harvest of cells, sub cellular components and macromolecules in high yields. It is commonly used as a separation technique when gravity separation is too slow and particles within the sample to be separated do not settle readily or at all. This technique has since paved the way for the determination and understanding of fundamental cellular processes such as oxidative phosphorylation, intracellular digestion, protein synthesis as well as transportation of substances in and out of the cell (Graham and Rickwood). Thus centrifuges are now regarded as one of the most basic yet valuable pieces of equipment required in biological laboratories.

This book aims to provide an introduction into the history and principles of centrifugation as well as an in depth explanation of the history and principles surrounding centrifugation. It also aims to provide an explanation of commonly used centrifugation techniques and their related practical applications in various industries. Furthermore, to aid in understanding the various principles behind this technique, the basic design of various rotors have been diagrammatically presented. Also in this book is an overview of the currently available centrifuge types and related instrument components as well as a description of centrifugal theory, a table comparing densities of commonly separated macromolecules and additional information on separation techniques together with a comprehensive advice for proper centrifuge handling, maintenance and related safety protocols. There is also further information on sample preparation and and how to design an ideal centrifugation protocol.

Author

CHAPTER 1

INTRODUCTION TO CENTRIFUGATION

Centrifugation is a standout among the most imperative and generally applied research methods in organic chemistry, cell and molecular science, assessment of suspensions and emulsions in drug store and in medicine. Present research and clinical implementations depend on the frequent segregation of cells, subcellular organelles and macromolecules, in exceptional quantities. A centrifuge utilizes divergent power (g-force) to separate suspended particles from their encompassing medium on a cluster or on a consistent – stream premis. (Rickwood & Graham, Biological Centrifugation, 2001). Numerous particles or cells in a fluid suspension will in the long run settle at the base of a compartment due to gravity. However the time span required for such detachments is often unreasonable. Different particles, to a small degree in measure, won't separate at all in the arrangemen, unless they are under high centrifugal force. The power exerted on the particles (contrasted with gravity) is called Relative Centrifugal Force (RCF). A common example would be where a Relative Centrifugal Force (RCF) of 500 x g demonstrates that the centrifugal force is actually 500 times more noteworthy than the natural gravitational power. Centrifugation is a procedure that includes the utilization of the centrifugal force for the sedimentation of heterogeneous blends around an axis of rotation. This technique is often utilized in industry and research based settings and involves a procedure that contains at least two immiscible fluids. When a centrifugation run occurs, progressively thick segments of the heterogenous blend will move far from the axis of the rotation, while less-thick segments of the blend will move towards the axis.

Drug specialists, physicists and scientists may reconstruct such a powerful gravitational force in a test tube in order to move quickly and fully influence the precipitate ("pellet") accumulation on the base of the tube. The rest of the arrangement is commonly called the "supernate" or "supernatant liquid". This supernatant liquid is then immediately emptied from the tube without aggravating the precipitate, or instead can be pulled back with a Pasteur pipette. (Rickwood, Ford, & Steensgaard, John Wiley & Son Ltd, 1994)

As articulated by Sir Isaac Newton in the first law of motion, a uninhibitedly moving body, (for example, a ball) tends to move in a straight direction, and if coordinated along a path of motion with some form of controlling power, (for example, a string held with a hand would be tied to it) it will exert some force against the coordinating or controlling power. This power, which is the outward draw of the ball on the string, is known as the centrifugal power. This power is measured by taking into consideration the precise speed of revolution, the mass of the object and the range of the circle through which the object moves. Multiplying the mass of the pivoting object will increment the centrifugal force 2 times more, yet multiplying the quantity of revolutions per minute (rpm) will expand the centrifugal force by 4 times (measures up to 2 times 2); likewise, expanding the speed by 10 times is going to increment the power by 100 times more (breaks even with 10 times 10). Centrifugal force can be deonstrated using the following equation:

$$F = mv2 / R = 4\pi 2mn2R$$

in which F is the centrifugal force, m is the mass, R is the range, v is the speed, n is the quantity of cycles occured every second. (Ford & Graham, 1991). A centrifuge itself is defined as any gadget that applies a supported centrifugal force; that is, a power exerted due to rotation. The centrifuge itself imitates a comparative, more grounded power which is present due to the power of gravity. Each centrifuge has a turning vessel; of which there are numerous designs contingent upon utilization. For example, a punctured pivoting drum that throws off abundant water from garments (a washing machine) is a sort of rotator. Once such industrial example is utilized as a part of the industry to isolate liquids after smashing.

The world's gravitational power is adequate enough to isolate many sorts of particles after some time. A container of anticoagulated blood left remaining on a seat top, in the end is going to separate out into plasma, red blood cells and white blood cells. however, the period of time necessary for such separations blocks this way of partition for general applications. Practically, centrifugal force is important to isolate general particles however apart from the time issue; the potential corruption of organic mixes during such delayed time capacities serves as a strong indicuation that quicker partitioning methods are required.

The rate of partition in a suspension of particles using the way of gravitational power basically relies upon the particle or molecule size and thickness. Particles of higher thickness or bigger size normally spin at a rate much faster and eventually will be isolated from particles of a reduced thickness. This sedimentation of particles, involving cells as well, can be clarified by the Stokes Equation.

Equation 1.1 computes the speed of sedimentation using five parameters. (Rickwood, Centrifugation: A Practical Approach, 1992)

$$V = \frac{d2(p-L)3g}{18n} \qquad (1.1)$$

in which:

v = sedimentation rate or speed of the circle

d = diameter of the circle

p = particle or molecule thickness

L = medium thickness

n = thickness of medium

g = gravitational power

Taking the parameters of Stokes Equation into consideration, five imperative properties of particles can be clarified:

- The rate of particle sedimentation is relative to the measurement of the molecule.
- The sedimentation rate is relative to the distinction in thickness between the molecule and the medium.
- The sedimentation rate is zero when the molecule thickness is the same as the medium thickness.

- The sedimentation rate diminishes as the medium consistency increases in increments.
- The sedimentation rate increases in increments as the gravitational power increases incremently.

Table 1: Centrifuge Classes and applications

Centrifuge Classes			
	Low – speed	High – speed	Ultra/micro - ultra
Maximum Speed (rpm x 103)	10	28	100 / 150
Maximum RCF (x103)	7	100	800 / 900
Pelleting Applications			
Bacteria	Yes	Yes	(Yes)
Animal and plant cells	Yes	Yes	(Yes)
Nuclei	Yes	Yes	(Yes)
Precipitates	Some	Most	(Yes)
Membrane fractions	Some	Some	Yes
Ribosomes / Polysomes	-	-	Yes
Macromolecules	-	-	Yes
Viruses	-	Most	Yes

Relative Centrifugal Force is reliant on the speed of rotation in revolutions per minute and the separation of the particles from the focal point of rotation. Whenever the speed of rotation is given in revolutions per minute (Q) and the distance (r) is communicated in centimeters, the Relative Centrifugal Force can be computed by utilizing the Equation for centrifugal force (RCF) 1.2.

$$RCF = 11.17 \times Rmax\ (RPM/1000) \quad (1.2)$$

WHAT IS CENTRIFUGATION?

A centrifuge is a device that separates particle from a solution through the use of a spinning rotor. Centrifuges are used to generate a centrifugal acceleration that is much greater than that of the Earth's gravitational force, thereby speeding up the sedimentation and purification of different components within heterogeneous mixtures. This separation is often based on the particles density and size and is most commonly used in industry and in laboratory settings. In addition, a range of other parameters are also known to affect centrifugation such as particle shape, choice and viscosity of medium for separation, and rotor speed. In biology, the particles to be separated are most commonly cells, sub cellular organelles or large molecules.

Over time, as advances in biological and technological research were made, such was the recognition of the importance of large molecules like proteins and nucleic acids. Through this came the need for developing new tools to study and analyse them. One of the most influential developments in the study of macromolecules was that of the ultracentrifuge by Svedberg and his colleagues in the 1920s (Svedberg and Pedersen, 1940). At this time, the opinion was that macromolecules did not exist; however, the pioneering studies of Svedberg led to the undeniable conclusion that proteins were truly macromolecules containing a huge number of atoms linked by covalent bonds. Later, substances such as rubber and polystyrene were shown to exist in solution as giant molecules whose molecular weight was independent of the solvent used. As continuous advances in the field of molecular biology were made, it then became possible to differentiate and separate the structures of biological molecules such as DNA and proteins.

Today there are a range of centrifuge designs and configurations available which are utilized in the mineral, petrochemical, chemical, medical, pharmaceutical, municipal waste, industrial waste, dairy, food, polymer, energy and agricultural industries.

There are various sorts of centrifuges, that can be arranged by expected utilization or by rotor outline. From the substantial floor assortment to the micro centrifuge, there are numerous assortments accessible for the specialist. These centrifuge categories include:

Benchtop Centrifuges: these are an expansive class of centrifuge named after their little seat space impression. Speeds reange in RCFs from as low as a couple of hundred to more than 50,000 x g. Tube volumes can extend from under 1 mL, (for example, with PCR tubes) to a couple of liters. Diverse sorts of rotors, for example, settled point, swinging container, and nonstop stream are additionally ordinarily compatible with benchtop models.

Refrigerated Benchtop Centrifuges are minimized instruments perfect for centrifugation of samples that might be temperature senstitive, for example, live cells or proteins. Many of these include exchangeable rotors and connectors to accomodate an extensive variety of test volumes from under 1 mL to a couple of liters. Velocities can also differ, and a few models can reach up to 60,000 x g.

Clinical Benchtop Centrifuges are simple, low-speed centrifuges that are perfect for the partition of entire blood segments, such as serum, plasma, red platelets, and other organic liquids. Their velocities may run between approximately 200 rpm to 6,000 rpm. Most clinical centrifuges can suit normal blood draw tubes, however communicating with the supplier is imperative to ensure availability of particular tube sizes or tube connectors.

Microcentrifuges are staple instruments in many research labs which for the most part utilise very small tube volumes such as 2 mL, 1.5 mL and 0.5 mL. Microcentrifuges tend to rotate at speeds of up to 16 000 x g, while more specific instruments can achieve acceleration rates of up to 30 000 x g. Likewise, producers may offer tradable rotors and tube connectors so users should be mindful of their requirements.

Vacuum Centrifuges or Concentrators may utilize vacuum, centrifugal power, temperature or potentially gas to evacuate fluid dissolvable for the concentration or dessication of tests. This instrument is perfect for cleansing samples such as nucleic acids, proteins, peptides, as well as different mixes of solutions.

THE HISTORY OF CENTRIFUGATION

A centrifuge within a lab forms an exteremly important piece of equipment that is extremely important and which is still being used today. It is thought to be one of the preferred techniques among the other productive approaches used for isolating samples of various densities. Many sorts of centrifuges are accessible with some bigger instruments being abale to achieve velocities of up to 10,000 rpm, while the little centrifuges as a rule have maximum rotational speeds underneath 300 rpm.

Seat top centrifuges equipped for paces of around 3,000 rpm have been being used since the mid-1800s. Early instruments were hand fueled, yet in 1912, with the onset of electric centrifuges, everything we since knew about these machines changed. The early centrifuges were generally utilized for non-organic applications, such as isolating milk. Centrifugation has been widely used in both industrial and laboratory settings for over hundreds of years. The earliest studies utilizing centrifugation techniques date back to the mid-15th century where centrifuges were used in the separation of milk. These machines were hand driven and could reach speeds of up to 3000 rpm (Wilson et al, 2009). Antonin Praddtl presented the first centrifuge for industrial purposes in 1864. This was also used for milk processing in order to separate cream from milk (Holtbrügge et al, 2016). Friedrich Miescher then expanded the potential of centrifuges to be used in laboratory settings in 1869. He used a crude system to isolate cellular organelles which eventually led to the discovery of a new class of biological components at the time, known as nucleic acids. Later, in 1878, Gustaf de Laval introduced the first continuous centrifuge for industrial applications thereby making commercialization of centrifuges possible and enabling the use of centrifuges in a wider range of applications (Wilson et al., 2009). By 1900, it was generally accepted that proteins were composed of amino acids; howeverllscientists were still unsuof whether proteins were colloids or macromolecules. During the 1920s and 1930s, Theodor Svedberg developed the first ultracentrifuge. His research group built a number of ultracentrifuges, which were then used to determine the molecular weights and subunit structures of highly complex proteins such as haemoglobin.

Through a series of experiments utilizing the sedimentation equilibrium technique, Svedberg and his group determined the molecular weights and subunit structure of haemoglobin, thereby proving that it was in fact a single compact particle. The results of these studies subsequently shaped new ideas about the structures of proteins. By 1923, Svedberg had developed an analytical centrifuge equipped with a photographic absorption system, which would exert a much greater centrifugal effect in order to analyse large grained sols. His research won him the Nobel Prize in 1926 where he was acknowledged for his work in colloid chemistry and the invention of the ultracentrifuge. His work has since paved the way for protein science. During this time Edward Pickels and Johannes Bauer built the first high speed vacuum centrifuge, which was used for studying viruses. In 1942, Albert Claude then initiated a quantitative fractionation procedure, which shifted the focus of centrifugation to analytical cell fractionation (Claude, 1946). His work on tissue fractionation outlined a procedure involving a series of centrifugation steps in which chromatin threads where eventually harvested from the final sediment. His work has since opened the way for the elucidation of fundamental cellular processes like oxidative phosphorylation, intracellular digestion, protein synthesis and bulk transport in and out of the cell (Volkl, 2010).

In 1946, Pickels cofounded the Specialized Instruments Corporation (Spinco) in Belmont, California and, while sales were not great in the first few years, things dramatically changed after the introduction of the first preparative ultracentrifuge in 1949. Beckman Instruments subsequently bought out SPINCO in 1954. Today they are known as Beckman Coulter and are still a major manufacturer of centrifuges. Netheler and Hinz Medizintechnik was a German based company, today known as Eppendorf, that subsequently introduced the first micro centrifuge for benchtop laboratory use in 1962. Highly sophisticated and versatile high speed centrifuges now form one of the most basic pieces of equipment required for everyday use in biological laboratories all around the worll Ccentrifugation has come a long way since the 1940's by moving away from rotors made of tensile steel to that of aluminium and titanium with changeable rotor systems offering standardized electronic features for cooling and imbalance detection.

In 2000 Eppendorf enhanced its line of great microcentrifuges together with the cooled 5415D centrifuge, that was actually the littlest and calmest cooled microcentrifuge available. Additionally in this year, Eppendorf propelled the MiniSpin and the MiniSpin Plus, known as individual centrifuges - a new epoch for the market of centrifuges.

Centrifuges in the Future

Centrifugation has made considerable progress since the procedure was first popularized for research center use in the 1940s. The first rotors, for example, those worked by Svedberg, were made of elastic steel. Today materials, for example, aluminum amalgams and titanium are utilized as a part of request to withstand high centrifugal forces. Standard elements now incorporate procedures for cooling, programming, programmed lopsidedness recognition, commotion decrease and alterable rotor frameworks.

Vacuum frameworks have additionally been added to current centrifuges to lessen erosion and keep up temperature control. The advancement of different instruments, for example, electron magnifying instruments, has enabled analysts to better inspect and examine the particles being centrifuged.

With look into proteins and cell nucleic matter winding up increasingly critical and picking up pace constantly, the centrifuge is going to keep on being an indispensable bit of lab gear for a long time to come. Advance from home, the Human Research Facility of the Destiny Laboratory of the International Space Station is as of now making utilization of centrifuge innovation from Hettich as a major aspect of its progressing push to enhance medicinal research and innovation for human life later on.

THE PRINCIPLES OF CENTRIFUGATION

Centrifuges have three essential parts that are described below in detail:

- A rotor
- A drive shaft
- An engine

The rotor holds the tubes, jugs, or sacks containing the fluids to be centrifuged. Distinctive rotor models and sizes which are often exchangeable with each other, can be mounted on the drive shaft, which interfaces with the engine. The engine possesses the ability to turn the rotor. Most centrifuges have a stopping mechanism to convey a message that will halt the rotor not long after the run is done. A lot of centrifuges additionally contain a refrigeration component to keep sensitive organic samples from getting warm and being subsequently destroyed.

Sedimentation is apparent in almost everything we do and forms a part of our everyday life. The effect of sedimentation is due to the influence of the Earths gravitational field ($g = 981$ cm.s^{-2}). However, this rate can be increased in a centrifugal field ($g > 981$ cm.s^{-2}) during experimental analysis. Many particles in a liquid suspension will eventually settle at the bottom of a container over time. Take for example a bucket of water that has sand particles added to it. The sand particles will travel slowly to the bottom of the bucket due to the earth's gravitational force. However, depending on the size and density of the particles, the length of time required for separation of certain particles may take longer than others. This is both time consuming and impractical. When that same bucket containing water and sand particles is swung around continuously, it now has what we call a centrifugal force exerted on it. This will, in turn, allow the sand particles to separate out and sediment at an increased rate. From this, we can deduce that particles subjected to a continuous centrifugal force under the influence of gravity will move towards the bottom of a vessel. However, the rate at which they move depends on their size and density.

Centrifugation is designed to use centrifugal forces, which are greater than the force of gravity, to speed up the sedimentation of particles. Therefore, when spinning a vessel with suspended particles around an axis of rotation at a certain given speed, a force is imposed on the particles causing them to move away from the axis of rotation. This force that is imposed is measured in multiples of the Earth's gravitational force and is known as the Relative Centrifugal Force (RCF), more commonly known as the *g* force. For example, an RCF of 600 x *g* indicates that the centrifugal force is 600 times greater than that of the Earth's gravitational force (Majekodunmi, 2015). The same principle can be applied to biological structures which may only have tiny differences in their size and density. The application of a relative centrifugal force provided by a centrifuge can be used to separate out various particles suspended in a solution thereby causing an overall increase in the rate of sedimentation. This is advantageous since there is a potential for degradation of biological compounds during prolonged storage further indicting the need for faster separation techniques. Particles whose densities are higher than that of the solvent in which it is suspended will sediment out faster while particles that are lighter in density will float to the top or require higher speeds and longer centrifugation times. Thus the greater the difference in density, the faster they will sediment out. If there is no difference in density then the particles will stay steady within the solvent. Therefore, it can be further deduced that the main purpose of applying a centrifugal force is to separate or sediment two or more immiscible liquids or particles contained in a heterogeneous mixture. Denser components will migrate away from the axis of rotation in the centrifuge and sediment at the bottom of the centrifuge tube while less-dense particles will migrate towards the axis. If the particles in a suspension are very small then high centrifugal effects will have to be applied to separate the particles. Thus, the actual size of the centrifuge machine is kept smaller but is rotated at very high speeds (rpm); whereas, if a large amount of particles is to be separated and a low centrifugal effect is sufficient, then the diameter of the centrifuge is increased while keeping the overall speed low. Particles that collect on the bottom of a tube are referred to as the precipitate or pellet, while the remaining solution is called the supernatant or supernatant liquid. The supernatant is then subsequently decanted from the centrifuge tube without displacing the precipitate and is generally removed using a Pasteur pipette (Rickwood et al. 1994).

A centrifuge can be used to separate a range of particles and macromolecules such as biological components in tissues and cells, subcellular components like cytoplasmic fluid, nuclei, mitochondria, Golgi bodies and proteins based on their density and nucleic acids like DNA and RNA (Table 2). Bench-top centrifuges can reach acceleration values of up to 15 000 *g* while high speed refrigerated centrifuges can reach speeds of 50 000 *g* and ultra-centrifuges which operate with refrigeration and a vacuum can reach up to 500 000 *g*.

Table 2: Densities of various biological materials and macromolecules (Griffith, 2010)

MATERIAL	DENSITY (g/cm^3)
Microbial cells	1.05 – 1.15
Mammalian cells	1.04 – 1.10
Organelles	1.10 – 1.60
Proteins	1.30
DNA	1.70

RNA	2.00
Golgi apparatus	1.06 – 1.10
Plasma membranes	1.16
Smooth endoplasmic reticulum	1.16
Mitochondria	1.19
Lysosomes	1.21
Peroxisomes	1.23
Soluble proteins	1.30
Plant viruses	1.30 – 1.45
Rhino- and enteroviruses	1.30 – 1.45
Nucleic acids	1.60 – 1.75
Ribosomes	1.60 – 1.75
Glycogen	1.70

DERIVATION OF THE PRINCIPLES OF CENTRIFUGATION

Centrifugation is used to separate particles based on their sedimentation properties which depend on a number of factors like size, density and shape. Particles that are 5 μm and larger in size will accumulate at the bottom due to the gravitational force but particles that are smaller than 5 μm can be described as moving with Brownian motion and, as a result, require a stronger centrifugal force for sedimentation. Density and shape will vary significantly depending on the composition of the solution in which the particles are suspended (Majekodunmi, 2015). Sir Isaac Newton noted in his first law of motion that a freely moving body, such as a ball for example, will tend to move in a straight line. If directed along a curved path by some restraining force like an attached hand-held string, it will exert an additional force against this restraining force. Thus it can be said that an object revolving in a circle exerts a force away from the centre of rotation. This force, which is the outward pull of the ball on its string, is known as the centrifugal force. For example, consider a body of mass (m) rotating in a circular path of radius (r) at a certain velocity (v). The force acting on the body in a radial direction can be given by the following equation:

$$F = \frac{mv^2}{r}$$

The gravitational force acting upon the same body can be given by the following equation:

G = mg

Where,

G = gravitational force

g = acceleration due to gravity

m = mass of body

The centrifugal effect is defined as the ratio between the centrifugal force and gravitational force. This can be described by the following equation:

$$C = \frac{F}{G}$$

$$= \frac{mv^2}{mgr}$$

$$= \frac{v^2}{gr}$$

Where,

F = centrifugal force

G = gravitational force

g = acceleration due to gravity

m = mass of body

v = velocity of the body

r = radius of circle of rotation

Since, $v = 2\pi r n$, where n = speed of rotation given in revolutions per minute (rpm),

$$C = \frac{F}{G}$$

$$= \frac{(2\pi^2 r^2 n^2)}{gr}$$

$$= \frac{4\pi^2 r^2 n^2}{Gr}$$

$$= \frac{2\pi^2}{g \quad Dr}$$

$= \text{kD n}^2$

Where, D = maximum diameter of the centrifuge and k is the constant defined as,

$$k = \frac{2\pi^2}{g}$$

D can be measured from the centre of the centrifuge to the free surface of the liquid or to the tip of the centrifuge tube. From the above equation ($C = kD\ n^2$) it can be deduced that: The centrifugal effect is proportional to the diameter of centrifuge, and The centrifugal effect is proportional to the speed of rotation

Therefore, from the above equation it can be deduced that by increasing the angular velocity of rotation, the mass of the object, or the radius of the circle through which the object moves, the amount of force can be increased too. In addition, while the centrifugal force is directly proportional to the radius and to the mass, it is proportional to the *square* of the angular velocity. For example, doubling the mass of the rotating object will increase the centrifugal force by a factor of 2 but doubling the number of revolutions per minute (rpm) will increase the centrifugal force by a factor of 4 (Majekodunm, 2015). Therefore, it can be said that if the particles in a solution are very small, then a high centrifugal effect will be required to separate the particles. The centrifuge size is usually kept smaller but is rotated at a very high speed, whereas if a large amount of material needs to be separated and a low centrifugal effect is sufficient, then the diameter (D) of the centrifuge is increased while the speed (n) is kept low (Figure 1).

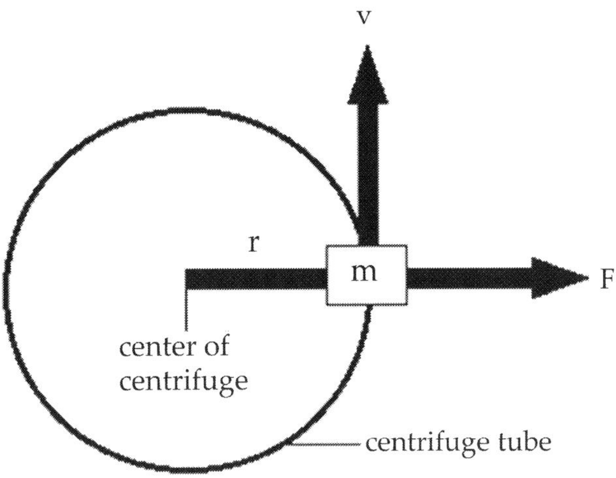

Figure 1: The basic principle of centrifugation (Majekodunmi, 2015).

Calculations of the relative centrifugal field are dependent on rotor speed, radius and duration of centrifugation. The rate of sedimentation is dependent upon the applied centrifugal field G (cms^{-2}), which is determined by the radial distance (r) of the particle from the axis of rotation (cm) and the square of the angular velocity (ω) of the rotor (in radians per second).

$$G = \omega^2 r$$

Where,

G = centrifugal field

r = radial distance

ω = angular velocity

The average angular velocity of a body that rotates about a fixed axis is defined as the ratio of the angular displacement in a given time interval and measured in radians per second. One radian is usually abbreviated as 1 rad. Since 360° equals 2π radians, one revolution of the rotor can be expressed as 2π rad. As such, the angular velocity in rads per second of the rotor can be expressed in terms of rotor speed *s* and shown as follows:

$$\omega = \frac{2\pi s}{60}$$

CHAPTER 2

THE THEORY OF SEDIMENTATION

Sedimentation is defined as a process whereby molecules or particles move and collect downward due to a gravitational force. The rate of separation in a suspension of particles depends on the particle size and density, whereby larger and denser particles will sediment faster than smaller and less dense particles. The process of centrifugation and application of a centrifugal force facilitates sedimentation which inevitably leads to the separation of particles into a solid pellet known as the precipitate and a liquid solvent known as the supernatant (Thomas, E., 2010).

Parameters that affect the sedimentation rate include the size and mass of particles, the buoyancy it is exposed to in the gradient medium and the frictional force which is influenced by the viscosity of the sample solution. There is a direct correlation between the rate of sedimentation and the size of particles to be separated, the speed of centrifugation as well as the density difference between liquid and solid phases of a solution (Beveridge, T, 2000). The interaction between these various forces during centrifugation is shown in Figure 2, which illustrates that the buoyancy and frictional forces affecting a particle counteract the centrifugal force. As a rotor spins in the centrifuge, each particle in the sample experiences a centrifugal force and will consequently sediment at a rate proportional to the force. At a fixed centrifugal force, the sedimentation rate of a particle is proportional to its size or molecular weight and to the difference between the particle density and the density of the solution (Griffith, 2010).

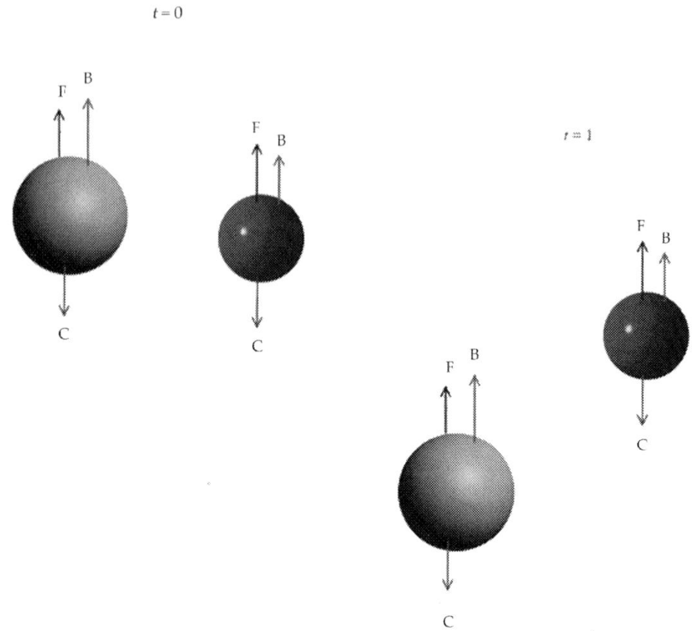

Figure 2: Schematic illustration of the interactions of various forces on the particles during centrifugation. These include the buoyancy (B), friction (F) and centrifugal (C) forces. (Griffith, 2010).

METHODS OF SEDIMENTATION ANALYSIS

There are two basic sedimentation strategies namely, vital or integral sedimentation and differential sedimentation. The accompanying discourse clarifies the contrasts between these strategies.

Integral Sedimentation

The Integral sedimentation strategy is the most established of all the sedimentation techniques and involves an indicator shaft (a light bar or X-ray pillar) that goes through the liquid at a known separation shape in the liquid surface, and measures the fixation of the particles.

The underlying power of light or that of the X-ray is basic and relates to the most extreme grouping of particles. While particles settle in between the liquid, the force of light that serving as a detection method of the particles now starts to increase.

Stokes' law can then be utilized to measure the particles that drag out of the liquid as a component of time, and an estimate of a molecule's conveyance is produced by plotting the deliberate centralization of particles against the ascertained molecule measurement.

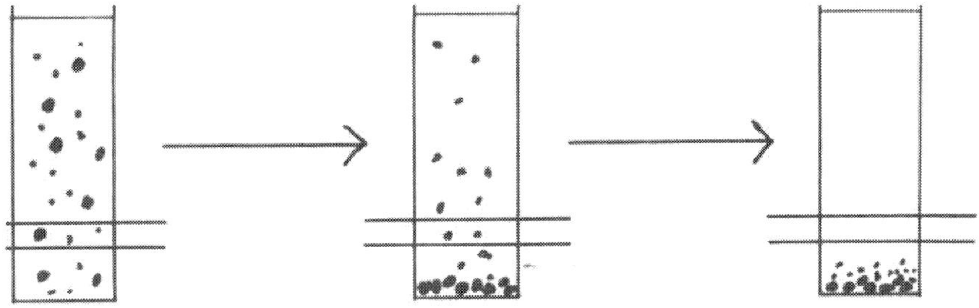

Figure 3: Integral Sedimentation Method.

There exist three huge operational issues with integral sedimentation that a centrifuge faces. To begin with, the underlying states of molecules during examination are hard to portray. In the event that the specimen is added to a centrifuge that is now turning, at that point there will be turbulent blending of the specimen as it scatters and is added to the axis. This could in turn influences the exact time estimation of sedimentation. In the event that a specimen is added to a centrifuge that is not turning but which later moves to a fast speed, at that point it is important to precisely quantify and represent the changing velocity amid the accelerating period. Likewise it is important to utilize a centrifuge that will ensure there is no blending of the specimen amid an increasing speed. Second, convection streams can be created in the time during an investigation unless the temperature of the specimen is held steady; any convection stream and flows in the liquid can decrease both determination and the preciseness of results. High speed centrifuges create frictional warmth, that makes it harder to keep up consistent temperature in the specimen liquid. Thirdly, the sedimentation chamber has to be purged and cleaned by following each specimen, which builds administrator work on a continual basis.

Differential Sedimentation

The method of differential sedimentation was discovered for the first time in the the 1900's. this method involves a specimen of particles that is to be broke down being put over a segment of clear fluid at the start of the examination. Particles settle as indicated by Stokes' Law, similarly as in the first method of integral sedimentation. Those particles with the highest weight will settle out first. The diminishment in force across particle size shows the centralization of particles in the indicator pillar. At the point when an X-ray bar is utilized, the lessening in force corresponds to molecule fixation.

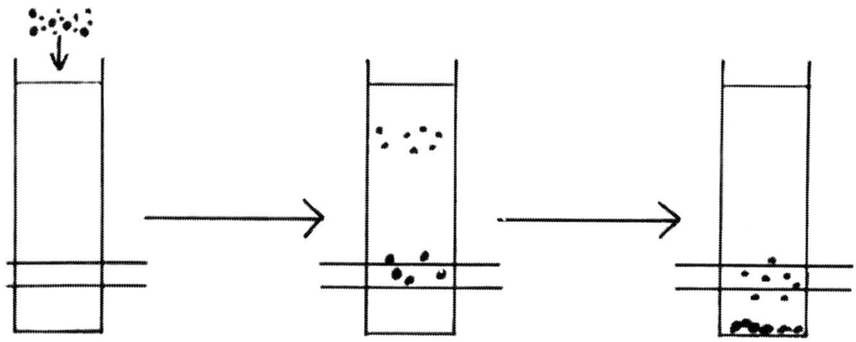

Figure 4: Differential Sedimentation Method.

At the point when all particles have passed the indicator, the flag comes back to the first level. A plot of the particle focus against the computed particle measurement creates a differential dispersion (shown below). The strategy is called differential sedimentation on the grounds that just a little piece of the appropriation (a "differential") is being measured by the indicator bar at any one time.

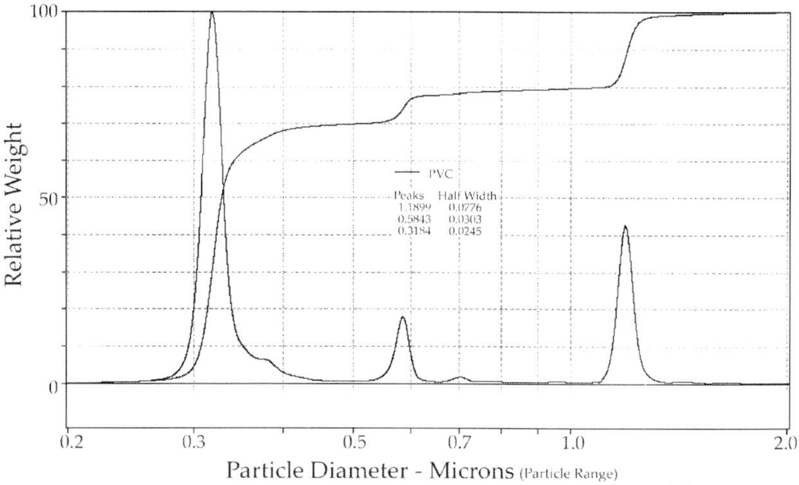

Figure 5: Differential and Integral Distributions.

Running a differential sedimentation is somewhat more difficult than proposed. At the point when a specimen of scattered particles known to be denser than the liquid itself in the section, is put over the segment, the particles don't settle exclusively as per Stokes' Law. Rather, the whole specimen suspension quickly settles as a mass liquid through the fluid segment, in the very same path as a homogeneous fluid of higher density (similar to 10% sodium chloride in water) would settle between a section of another fluid of lower density (for example water). The mass settling of a specimen in differential sedimentation is ordinarily called "streaming" or better known as "sedimentation instability" (Allen, 1968). All data about a particle's measurement circulation can be lost in the moment that streaming happens. Several types of methods have been produced to minimize and even prevent streaming. All of these strategies are compelling in light of the fact that a slight denseness of the slope is shaped inside the liquid section, before beginning examinations. An extensive variety of liquids may be utilized and these include methanol, ethanol, glycerine and sucrose. In nonaqueous frameworks, numerous blends of liquids of various densities can be utilized.

A density gradient takes out spilling on the grounds that constantly amid the run, the net density of the liquid, which is the normal density of liquid in addition to any suspended particles, increases in increments consistently throughout the liquid segment. The condition that ensures stable sedimentation is given by the following equation.

$\delta\rho net/\delta R \geq 0$

in which: ρnet is the net liquid density (counting fluid in addition to any suspended particles) and R is the separation from the focal point of pivot.

At the point when a little volume of a particle suspension is put on the surface of the liquid segment, the net density of the suspension is somewhat higher than the unadulterated liquid; however the liquid simply under the surface is additionally marginally higher in density than the unadulterated liquid, because of the density gradient or slope. There is no main impetus for mass settling of the molecule suspension, which in turn means there is no unsteadiness, and the particles residue through the liquid as indicated by Stokes› Law. The required steepness of the density gradient relies on the net thickness of the specimen to be measured. A specimen with higher net density) requires a more extreme density gradient than a specimen with low net density. Most specimens are weakened to low focus, so just an exceptionally slight density gradient is required to guarantee security. Density inclinations of under 0.01 g/ml per centimetre of liquid length are typically adequate.

A density gradient additionally kills warm convection, and in this way sedimentation is not affected by slight differences in liquid temperature during examination. Moderately huge temperature changes (greater than 0.5°C) can cause some loss of precision unless they are represented, in light of the fact that liquid density may change in accordance with temperature.

Differential Centrifugal Sedimentation Instrument Architecture

The most widely recognized architecture for the Differential Centrifugal Sedimentation instruments is an empty, optically clear circle that is driven by a variable speed engine. An ordinary plate cross segment is shown in the following figure. The plate can be for all intentions and purposes any size, yet manufacturers have settled on a measurement of roughly 1.25 to 1.50 cm. The indicator pillar is typically monochromatic light of generally short wavelength (400 nm - 500 nm); however a few instruments utilize a more drawn out wavelength (~650 nm), or X-rays. Shorter wavelength light gives better indicator affect-ability when particles smaller than 100 nm are measured.

To set up the instrument for examination, the plate is placed under way at consistent speed, and after that the circular chamber is loaded with a liquid that holds a slight density gradient. Tests are set up for investigation. The lower density liquid utilized for the specimen lowers any initial blending of the liquid inside the plate with the specimen. At the point when a specimen is infused (typically utilizing a little syringe), it strikes the back inner section of the circle, and structures a thin film, which spreads as it quickens radially toward the surface of the liquid. At the point when the specimen scattering reaches the liquid surface, it rapidly spreads over the surface, since it is of lower density which in turn means it can "drift" on the higher density liquid. Once a specimen is on the liquid surface, sedimentation of individual particles begin. The infusion of a specimen is quick (normally less than 50 milliseconds), so the start time for an examination is all round characterized and the accuracy of sedimentation time is very good.

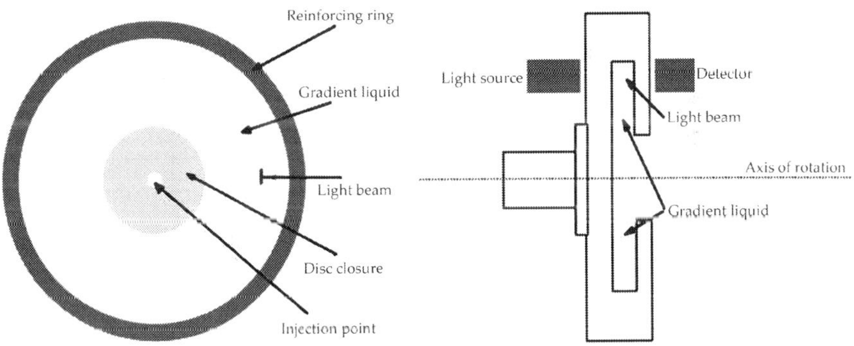

Figure 6: Hollow Disc Centrifuge Architecture.

At the point when an examination is finished, the instrument is prepared for the next specimen. There is no compelling reason to avoid immediate cleaning of the axis. The main restriction on consistent run time is that the density gradient gradually disintegrates because of atomic dissemination. At the point when the density gradient cannot keep up stable sedimentation, the instrument must be halted, exhausted, and another inclination framed. Usually the gradient lifetime is 2 to 72 hours, contingent upon the atomic weight and viscosities of the materials that shape the gradient.

Advantages and Limitations of the Differential Centrifugal Sedimentation Method

All strategies for particle separation may be described by three parameters:

- the accuracy of the detailed size conveyance;
- the repeatability of the revealed estimate appropriation;
- the determination of the appropriation.

Precision and Repeatability

Precision and repeatability of the differential centrifugal sedimentation technique is high in almost all cases. Any noteworthy mistakes in the outcome usually may have occured due to either changes in the physical parameters of the framework, shakiness in the sedimentation, or by deviation of the sedimentation from Stokes' Law.

Physical Parameters

The general accuracy of the investigation relies on the overall precision of each of the qualities in the equation below.

$\delta\rho net/\delta R \geq 0$

For instance, if the consistency of the liquid is really 2 percent higher than that which is entered in the equation above, at that point the revealed particle size will be around 1 percent less than that which is expected. It is conceivable to accomplish almost any coveted level of accuracy by enhancing the precision of the parameters in the equation above. An options to enhance precision is to utilize a tight adjustment standard of known size.

Sedimentation Stability

Any additional movements (streaming or shakiness) amid a centrifugation run decreases both the precision and determination of the run itself. To check that the sedimentation is steady, an immediate method for affirming soundness of the pellets is required. Some differential centrifugation sedimentation instruments contain a strobe light which is synchronized with the revolution of the axis. This allows for the visual coordination of sedimentation steadiness. In addition, different instruments depend on a thin adjustment standard to check dependability. At the point when an adjustment standard is utilized, assessment of sedimentation security can be programmed such that the width and state align with the width and shape of a known standard. Any noteworthy change in the width or shape will indicate precariousness in the sedimentation.

STOKES LAW

The basic principles of the sedimentation theory originated from Stokes Law. This law involves the sedimentation of a particle in a fluid and the resultant force which resists its motion through the fluid. This force is known as the frictional force (F). The Stokes Law Equation can be derived from the settling of a sphere in a gravitational field. As the velocity of the sphere reaches a constant value, the net force of the sphere is equal to the force resisting its motion through the liquid. This resisting force is known as the frictional or drag force.

$$(1/6)\ \pi\ d^3\ (p_p - p_l)\ g = 3\ \pi\ d\ \mu\ v$$

Where,

v = sedimentation rate or the velocity of a sphere

d = diameter of the sphere

p_p = density of the sphere

p_l = density of the liquid

μ = viscosity of the liquid medium

g = gravitational force

By solving for the velocity of the sphere, we get:

$$V = \frac{d^2(p_p - p_1)}{18\mu} \times g$$

From the Stokes equation, it can be seen that there are five important parameters that affect centrifugation. They are: the size of the particles; the density between the particles and the solvent in which it is suspended; the density of the solvent; the viscosity of the solvent and the gravitational force (Rickwood, 1992). From the above equation it can also be deduced that the rate of particle sedimentation is proportional to the size of the particles; the sedimentation rate is proportional to the difference in density between the particles and the solvent in which it is suspended; the sedimentation rate is zero when the density of the particles is the same as the density of the solvent; the rate of sedimentation decreases as the viscosity of the solvent increases and lastly the rate of sedimentation increases as the gravitational force increases. Therefore, it can be said that the process of centrifugation and subsequent particle sedimentation depends on several factors such as: size of particles, speed of centrifugation, density differences between the liquid and solid phases of a solution as well as the fluid viscosity (Bever 2000).

The Stokes Law equation can be written down as follows:

$$v = [d^2 (p_p - p_1) / 18\mu] \times g$$

$$= d_r / d_t \times g\ (981)$$

Where,

dr/dt = rate of movement of the particles in cm/sec

The sedimentation rate (v) can be shown by the following equation:

$$v = dr / dt$$

$$= s \times \omega^2 r$$

Where,

s = sedimentation coefficient (seconds)

r = distance between the particle and the centre of rotation (cm)

ω = angular velocity (radians / sec)

dr/dt = rate of movement of the particles (cm/sec)

The sedimentation coefficient is defined as the sedimentation rate per unit of centrifugal force and can be written as follows:

$$s = (1/\omega^2 r) \times dr/dt$$

The sedimentation coefficients relate to the molecular weight (M). Using the previous two equations we can deduce:

$$v / \omega^2 r = \frac{M(1 - Vp_1)}{\int}$$

Where,

M = mass or molecular weight of one particle

$V = 1/p_p$

$\int$ = molar friction constant

Continuing with the derivation, we can say:

$$S = \frac{M(1 - Vp)}{N \int}$$

Where,

M = mass or molecular weight of one particle

$V = 1/p_p$

$\int$ = molar friction constant

N = Avogadro's number

Taking into consideration the sedimentation and diffusion coefficient, Svedberg's equation can now be written as follows:

$$s / D = \frac{M(1 - Vp)}{RT}$$

Where,

D = diffusion coefficient

R = universal gas constant

T = absolute temperature

From the above equation we can see that the mass of a particle is proportional to its sedimentation coefficient and inversely proportional to its diffusion coefficient.

SEDIMENTATION OF PARTICLES IN A GRAVITATIONAL FIELD

When a solute particle is suspended in a solvent and subjected to a gravitational field, three forces are known to act on the particle (Figure 7). First, there is a sedimentation or gravitational force (Fs), which is proportional to the mass of the particle and the acceleration. In a spinning rotor, the acceleration is determined by the distance of the particle from the axis of rotation (*r*), and the square angular velocity (ω).

$$Fs = m\, \omega^2 r$$

$$= (M / N)\, \omega^2 r$$

Where,

m is the mass in grams of a single particle

M is the molar weight of the solute in g/mol

N is Avogadro's number.

Secondly, there is a buoyant force (F_b) that is equal to the weight of the fluid displaced:

$F_b = -m_0\, \omega^2\, r$

Where,

m_0 is the mass of fluid displaced by the particle

$m_0 = m\bar{U}\rho = (M/ N)\, \bar{U}\rho$

Where,

$\bar{U}$ is the volume (in ml) that each gram of the solute occupies in solution

ρ is the density of the solvent in g/ml.

If the density of the particle is greater than the density of the solvent, then the particle will begin to sediment. As the rotor spins and the particle begins to move along a radial path towards the bottom of the tube, its velocity (u) will increase because of the increasing radial distance. Since particles moving through a viscous fluid experience a frictional pull that is proportional to the velocity, the particles will then experience a frictional force which an be shown with the following equation:

$F_f = -fu$

Where,

f = the frictional coefficient (This depends on the shape and size of the particle)

Bulky or elongated particles experience more frictional drag than compact, smooth and spherical particles. The negative signs in the above equations indicate that these two forces act in the opposite direction to that of sedimentation.

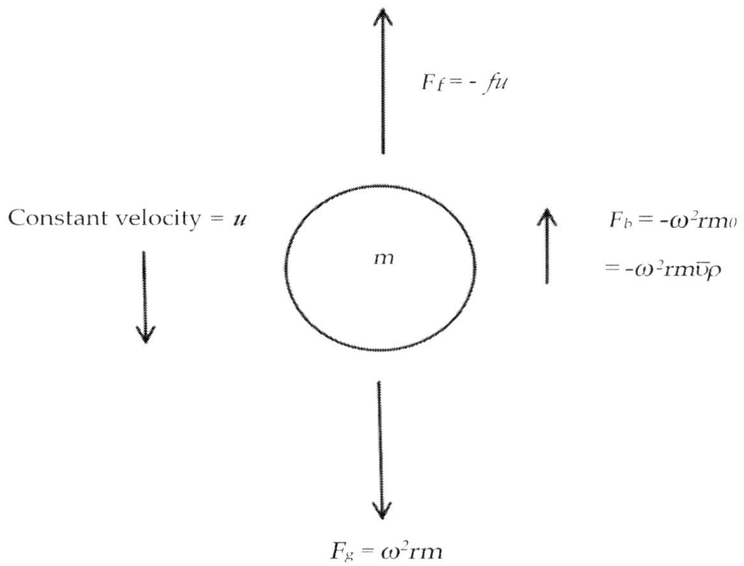

Figure 7: A solute particle is suspended in a solvent and subjected to a gravitational field with three forces acting on it. These forces are the buoyant force (Fb), the sedimentation force (and Fs) and the frictional force (Ff). (Ford & Graham, 1991).

During centrifugation, the three forces will come into balance. This can be shown as follows:

$F_s + F_b + F_f = 0 \omega^2 r \quad (M/N)\, v\rho\omega^2 r - fu \quad 0$

$(M / N)(1 - v\rho)\, \omega^2 r - fu = 0$

ThereforeTherefore, we can say,

$[M(1 - v\rho)] / Nf = u / (\omega^2 r) \equiv s$

Where,

$u / (\omega^2 r)$ is referred to as the velocity of the particle per unit of gravitational acceleration or more commonly known as the sedimentation coefficient.

From the above equation, it can be seen that the sedimentation coefficient depends on the properties of the particle. It is proportional to the molecular weight of the particle and inversely proportional to the frictional coefficient. In addition, it is independent of the operating conditions. Within a given centrifugal field, molecules with different molecular weights, sizes and shapes will move with different velocities and thus have different sedimentation coefficients. During sedimentation the solute will begin to collec at the bottom of the centrifuge tube and as the concentration at the bottom begins to increase, the process of diffusion will oppose that of sedimentation. After an appropriate period of time, the two opposing processes will approach equilibrium in all parts of the solution and for a single, ideal solute particle the concentration will increase exponentially towards the bottom of the centrifugation tube. At sedimentation equilibrium, the processes of sedimentation and diffusion are balanced, so the concentration distribution from the top of the cell to the bottom no longer changes with time and becomes a function of molecular weight. If the density of the particle is greater than that of the solvent, then the particle will sediment towards the bottom. However, if the density of the particle is less than that of the solvent, the particle will float towards the top of the solution (Figure 8). This is the situation for many biological structures such as lipoproteins and lipids in aqueous solutions. When the densities of the solute and solvent are equal then there will be no tendency to move in either direction and it is said that equilibrium is reached.

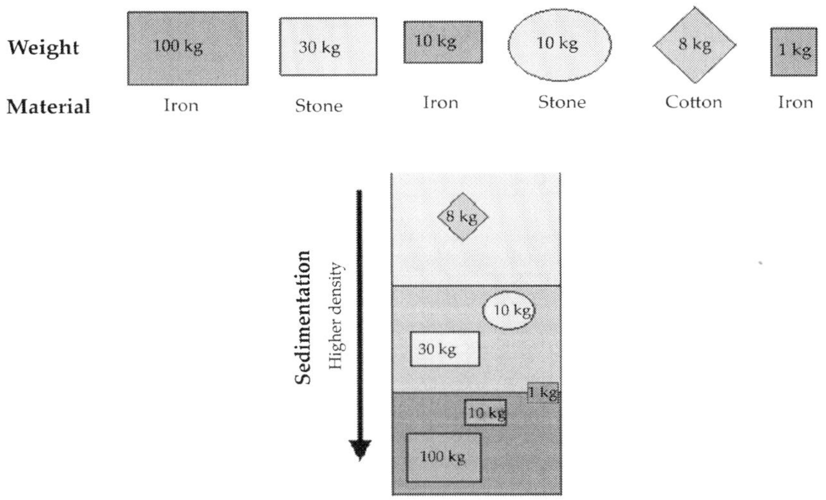

Figure 8: Separation of biological components by sedimentation showing that particles with a density higher than that of the solvent will sediment out to the bottom whilest those with a density less than that of the solvent will float towards the top.

One of the major uses of centrifugation is to determine the density of a macromolecule using a technique known as density gradient sedimentation. A gradient of density is most commonly made by generating a gradient spanning a range of concentrations. Commonly used solutes are sucrose or caesium chloride which are added in high concentrations at the bottom of the centrifuge tube and continue to be added in layers with successive lower values until reaching the top. The macromolecules to be separated will then sediment only when in a region of solution where the density is less than its own (Figure 8). When macromolecules are in a region of higher density then they will begin to float until eventually they form a layer at that region where the density of the solvent is equal to the density of their own. This is referred to as the buoyant density.

CALCULATING SEPARATION TIME

The separation time (T) is the time required for particles scattered in the solution to lower from the top (r_t) of the tube to the bottom (r_b) of the tube (Hitachi, 2013). Sedimentation is usually expressed as the sedimentation coefficient (s) which usually becomes a small value of s x 10^3. Separation time can be calculated using the following equation:

$$T\,(hr) = \frac{10^{13}}{3600 \times s} \times \frac{In\,(r_b) - In\,(r_t)}{\omega^2}$$

Where,

ω = angular velocity = $[(2\pi)/60] \times N$

N = speed (rpm)

r_t = distance from the axis of rotation to the top of solution

r_b = distance from the axis of rotation to bottom of the tube

If it is defined that,

$$K = \frac{In\,(r_b) - In\,(r_t)}{\omega^2} \times \frac{10^{13}}{3600}$$

Then,

$$T(hr) = \frac{K}{S}$$

If S is unknown, then the following formula can be used for calculation:

$$S = \frac{d^2(p_2 - p_1)}{18n \times 10^{13}}$$

Where,

d = diameter of lowering particle (cm)

p1 = density of solution surrounding particles (g/ml)

p2 = density of lowering particle (g/ml)

n = viscosity of solution surrounding particles (poise)

Note: Poise is defined as a non-SI unit of dynamic viscosity in the centimetre-gram-second system of units. So 1 poise = 1 gram per centimetre per second (1 g/cm/s). The viscosity of water at 20 °C is 0.01 Poise, S in pure water, will be expressed as $S_{20,w}$

CHAPTER 3

THE K FACTOR

The K factor (K), or clearing factor, varies depending on the type of rotor and the speed. When the rotor is used at a certain speed, the K factor (K_N) can be obtained from the formula below:

$K_N = K\,(N_{max} / N)^2$

Where,

N = speed used (rpm)

N_{max} = maximum speed (rpm)

K_N = K factor at the speed used

K = K factor at the maximum speed

The k factor is the relative pelleting efficiency of a given centrifuge rotor at maximum rotational speed. It can be used to estimate the time (t) in hours for sedimentation of a fraction with a known sedimentation coefficient (s).

$$t = \frac{k}{s}$$

The value of the clearing factor depends on the maximum angular velocity (ω) of a centrifuge (rad/s) and the minimum and maximum radius (r) of the rotor.

$k = [\ln (r_{max} / r_{min}) / \omega^2] \times 10^{13} / 3600$

Where,

r_{max} = maximum radial distance measured to the bottom of the tube

r_{min} = radial distance measured to the meniscus at the top of the tube

ω = angular velocity in radians / sec which is given as: $2\pi rpm / 60$ or $0.10472 \times rpm$

Since the rotational speed of a centrifuge is usually specified in RPM, the following formula is used:

$$k = [2.53 \times 10^5 \times \ln(r_{max} / r_{min})] / (RPM / 1000)^2$$

The k factor is related to the sedimentation coefficient (s) by the following formula:

$$T = K / S$$

Where,

T = time to pellet a certain particle (hours)

S = sedimentation coefficient

Since S is constant for a certain particle, the relationship can be interconverted between different rotors. Therefore, if the time to pellet a sample in an old rotor is known, then one can determine the time it would take to pellet a sample in a new rotor by using the following equation:

$$\frac{T1}{K1} = \frac{T2}{K2}$$

Therefore, to determine T1, we can say:

$$T1 = T2\,(K1 / K2)$$

Where,

T_1 = time to pellet in one rotor

K_1 = k-factor of the rotor

K_2 = k-factor of the other rotor

T_2 = time taken to pellet in the other rotor

New k factors can be calculated for a rotor running at a speed other than the maximum using the following formula:

$$K_{new} = \frac{(\text{maximum rated speed of the rotor})^2}{(\text{actual run speed of the rotor})^2}$$

Where,

K_{new} = new k factor for a rotor at a speed other than the maximum speed

K_{max} = k factor at the maximum speed of the rotor

To calculate the run time for a new rotor, the following equation can be used:

$$\frac{T(a)}{T(b)} = \frac{K(a)}{K(b)}$$

Where,

T(a) = time to pellet in the new rotor

T(b) = time to pellet in the old rotor

K(a) = k factor of the new rotor

K(b) = k factor of the old rotor

To calculate the run time for a new run, the following equation can be used:

$$T = \frac{(a \times b)}{c}$$

Where,

a = run time from a previous study

b = *g* force of a rotor for a previous study

c = *g* force for the new study

USING THE K FACTOR TO COMPARE ROTOR EFFICIENCY

While most researchers rely on the maximum speed to determine rotor efficiency, it should be kept in mind that a range of other factors like the centrifuge tube geometry and temperature also contribute to rotor performance and should be considered prior to making a valid and final comparison between rotors. Previous studies have shown that some rotors are more efficient at slower speeds than comparable rotors at higher speeds due to the influence of these other factorsreTherefore, using the k factor to compare rotors rather than the maximum speed alone is much more accurate.

Separation is affected by the maximum speed, maximum radius (r_{max}) and minimum radius (r_{min}) which together will determine the maximum *g* forer; howeve,, the particle path length also affects the separation time. The greater the minimum radius, the greater the centrifugal force at the top of the tube and the faster the separation will be. In addition, the centrifugal force at the minimum radius will define the minimum size that a particle needs to be in order to start sedimenting. In addition, the total path length of the rotor (also known as the difference between the maximum and minimum radius) is also known to affect rotor efficiency. Thus it can be said that the *k* factor is a simple measure of the overall rotor efficiency which incorporates both *g* force and particle path length. The diameter of the sample tube and the angle at which the tube is held will directly affect the path length. For example, a centrifuge tube that is more sharply angled will result in a shorter path length. A previous study conducted by Beckman Coulter (2014) to determine the efficiency of a Beckman Coulter Type 70.1 Ti and Type 70 Ti rotor showed that the initial rotor when run at a lower speed will pellet material faster than the later which was run at a higher speed. The Type 70.1 Ti rotor was run at 450 000 x g while the Type 70 Ti was run at a higher RCF of 504 000 x g at the same rotations of 70 000 rpm. The tube cavity of the Type 70.1 Ti has a smaller r_{max} and path length which in turn was more favourable and resulted in over 18 % faster pelleting time than in the Type 70 Ti.

To calculate the overall rotor efficiency, the following equation can be used:

$$k = [2.533 \times 10^5 \times \ln(r_{max}/r_{min})] / (RPM / 1000)^2$$

RELATIVE CENTRIFUGAL FIELD

The term centrifugal force is derived from the Latin word *centrum* meaning *centre* and *fugere* meaning *to flee*. It is defined as the apparent force that draws a rotating body away from the centre of rotation and is caused by the inertia of a body as the body's path is continually redirected. There are a range of particles that can be separated based on their size, shape and density using centrifugation. These include cells, proteins and nucleic acids. The rate of separation depends on the particle size and density. Particles of a larger size and consequently higher density will travel at a faster rate than those particles that are smaller and less dense. Most particles are so small that the effect of a gravitational force is insufficient to cause separation. Thus using the technique of centrifugation is one such way to increase the magnitude of the gravitational field and subsequent separation of different particles in a solution. This applied radial centrifugal force then moves them away from the axis of rotation. Thus centrifugal force can be defined as the apparent outward force that drives a rotating body away from the centre of rotation usually caused by the inertia of the body as the body's path is continually redirected. Relative centrifugal force generated by a rotor depends on the actual speed of the rotor in revolutions per minute (rpm) and the radius of rotation, also known as the distance from the axis of rotation (Beckman, 2005). The RFC (Relative centrifugal field) depicts the quality of the fields produced by various size rotors and working velocities. Similarly as length is measured in units of inches or millimeters, time in units of hours or minutes, the relative radiating field is measured in RCF and is communicated in gravitational fields, g. The relative centrifugal field differs with the square of the rotor speed. This in turn means that any adjustment in speed will cause a considerably more prominent change in the relative centrifugal field.

Derivation of Relative Centrifugal Force

The force acting on a particle during centrifugation can be given by the following equation:

$$F = m\omega^2 r$$

Where,

m = mass of the particle

ω = angular velocity of the spinning rotor in radians per second

r = distance of the particle from the axis of rotation

Relative centrifugal force is defined by the following equation:

$$RCF \cdots = \frac{F_{centrifugation}}{F_{gravity}}$$

$$= \frac{m\omega^2 r}{mg}$$

$$= \frac{\omega^2 r}{g}$$

Where,

g = acceleration due to gravity which is 980 cm / sec / sec

m = mass of the particle

ω = angular velocity of the spinning rotor in radians per second

r = distance of the particle from the axis of rotation

Since,

$\omega = [(2\pi rpm) / 60]^2$ and g is a constant we can therefore say,

$RCF = [(2\pi rpm) / 60]^2 r \times g^{-1}$

$= [(2\pi rpm) / 60]^2 r \times (980)^{-1}$

$= 1.12 \text{ x} 10^{-5} \text{ rpm}^2 \times r$

Therefore, to determine the RCF during a centrifugation run, it is necessary to measure the revolutions per minute (rpm) of the rotor and the distance between the sample and the axis of rotation. Since the RCF is defined as being the ratio of the two forces, it has no units; however, it is usually followed with a symbol g which indicates that the applied RCF is a multiple of the earth's gravitational force.

How to Calculate Relative Centrifugal Field

In centrifugation, it is important to differentiate between the speed of centrifugation given in revolutions per minute (RPM) and the relative centrifugal force (RCF or g) as these are most commonly confused (Rickwood, D, 2014). Revolutions per minute (RPM) is not a valid unit because the centrifugal force depends on the radius of the centriever; however, if the RPM and the radius of the rotor is known then the RCF can be calculated.

The centrifugal field is usually expressed in multiples of the gravitational field g. The relative centrifugal field (RCF) can be calculated using the following equation:

$$RCF = \frac{4\pi 2(\text{rev min} - 1)2r}{3600 \times 981}$$

$$= \frac{G}{g}$$

The equations that are used to calculate RCF from a known RPM and radius of rotation as well as calculate RPM from a known RCF and radius are interchangeable and can shown as follows:

Calculations used to convert RPM to RCF:

$RCF = 1.12 \times r\ (rpm / 1000)^2$

Calculations used to convert RCF to RPM:

$RPM = [\sqrt{(RCF/ (r \times 1.12)}] \times 1000$

Where,

r = the radius or distance from the centre of rotation in centimetres

From the above equations, it can be seen that the centrifugal force increases as the particles move down the centrifuge tube and the greater the centrifugal force, the shorter the separation time (Rickwood, 2014).

Calculation Examples

Example 1: Calculation of centrifugal field

Question: What is the applied centrifugal field at a point equivalent to 5cm from the centre of rotation with an angular velocity of 3000 rad s^{-1}?

Answer: The centrifugal field, *G*, at a point 5cm from the centre of rotation may be calculated using the equation:

$G = \omega^2 r$

$= (3000)^2 \times 5\ cm\ s^{-2}$

$= 4.5 \times 10^7\ cm\ s^{-2}$

Example 2:

A fixed angle rotor exhibits a minimum radius r_{min}, at the top of the centrifuge tube of 3.5 cm and a maximum radius r_{max} at the bottom of the tube of 7.0 cm (Figure 9). If the rotor is operated at a speed of 20 000 rpm, what is the relative centrifugal field (RCF) at the top and bottom of the centrifuge tube?

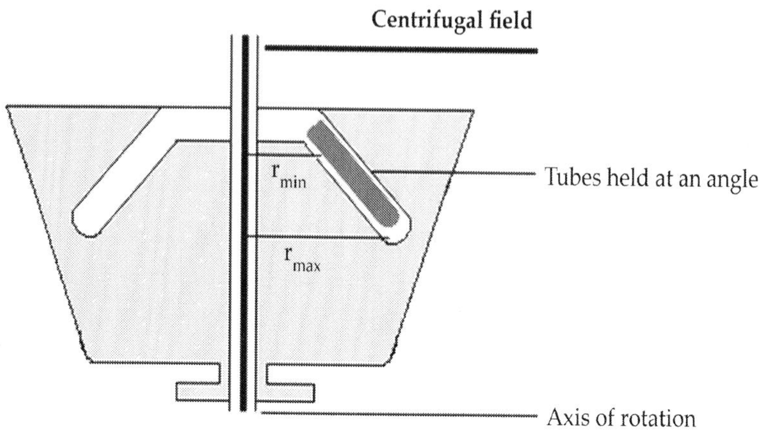

Figure 9: Schematic diagram illustrating the minimum and maximum radius of a fixed angle rotor.

Answer: The relative centrifugal field may be calculated using the following equation:

$RCF = 1.12 \times 10^{-5}\ rpm^2 \times r$

RCF at the top of the centrifuge tube:

$RCF = 1.12 \times 10^{-5} \times (20\ 000)^2 \times 3.5\ g$

$= 15\ 680\ g$

RCF at the bottom of the centrifuge tube:

$RCF = 1.12\ x10^{-5} \times (20\ 000)^2 \times 7.0\ g$

= 31 ereforeTherefore, it can be seen that with fixed angle rotors, the centrifugal field at the top and bottom of the centrifuge tube may differ. In this case that difference is two-fold. It should be noted that the value of the radius will depend on the shape and type of rotor within the centrifuge. If a fixed angle rotor is used as in above, there are usually two different values that need to be used for the radius. However, with a swinging bucket rotor for example, the tubes start off initially hanging in the rotor and as the rotational speed increases, so does the radius (Figure 10 and 11).

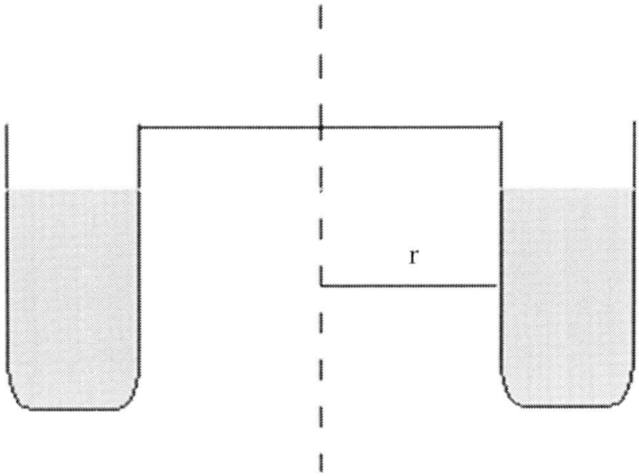

Figure 10: Schematic diagram showing the initial angle of centrifuge tubes when held in a swinging bucket rotor prior to a centrifuge run.

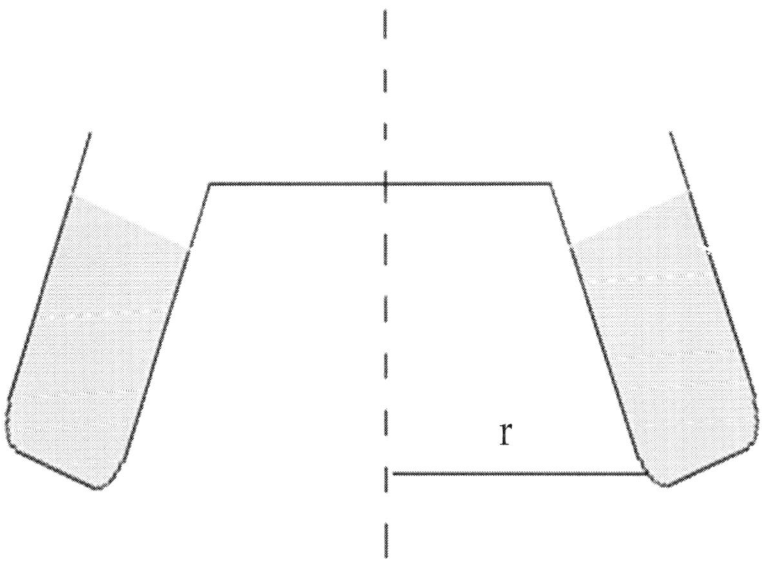

Figure 11: Schematic diagram showing the angle of centrifuge tubes in a swinging bucket rotor during a centrifuge run. As the rotational speed increases, the radius also increases.

Example 3: Calculation of angular velocity

Question: An ultracentrifuge is operated at a speed of 40 000 rpm. Calculate the angular velocity, ω, in radians per second.

Answer: The angular velocity, ω, may be calculated using the equation,

$\omega = [2\pi \text{ rev min}^{-1}] / 60$

$= 2 \times 3.1416 \times 40\,000 / 60 \text{ rad s}^{-1}$

$= 4188.8 \text{ rad s}^{-1}$

Calculation of Relative Centrifugal Field using a Nomogram

Although RCF can easily be calculated, centrifugation manuals usually contain a graph called a nomogram (Figure 12) which ows forthe easy extrapolation and conversion between relative centrifugal force and speed of the centrifuge. By lining up a known rotating radius (the distance from the centre of the rotor to the bottom of the centrifuge bucket) on the left with the known rpm value of the far right, the RCF value can be read off the nomogram at the central point where the line crosses the given graphs. Most manufacturers will include a nomogram in the instrument manuals which allow users to calculate RCF and RPM but many modern centrifuges also have the ability to swap the figures displayed on the control panel between RPM and RCF thereby making manual calculations unnecessary.

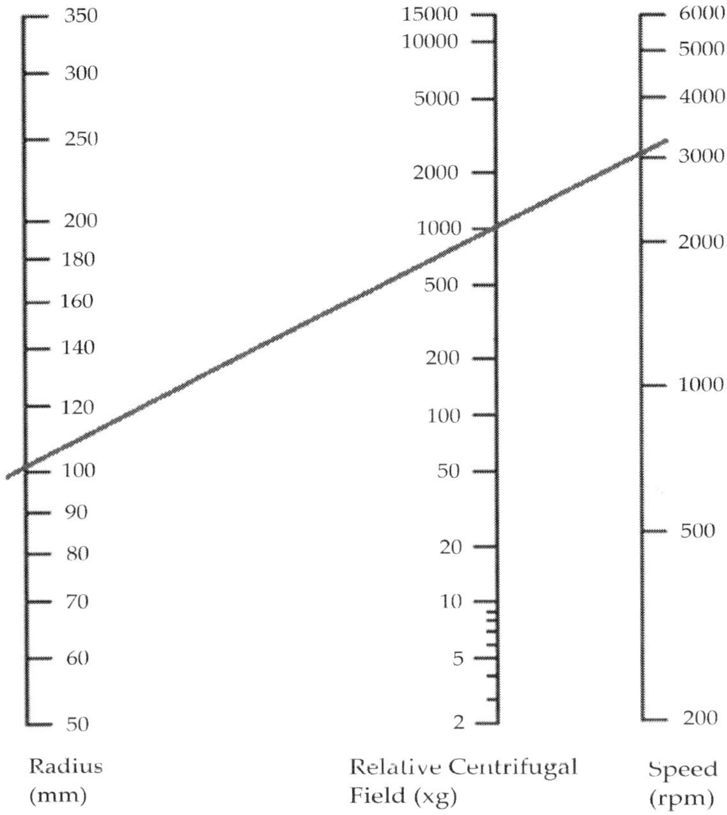

Figure 12: Nomogram for a tabletop rotor used to convert revolutions per minute (RPM) to relative centrifugal field (RCF) and vice versa.

To find the RCF at a radial distance of 100 mm from the centre of rotation when operating the centrifuge speed at 3000 rpm, place a straight line on the chart connecting the mm point on the Rotating Radius scale with the rpm point on the Speed Scale. Read the point at which the straight line intersects with the Relative Centrifugal Force Scale. In this case the RCF is 1000 x *g*. Similarly if the RCF is known, the speed for a given rotor can be calculated by connecting the points and reading the intersection on the Speed scale.

CHAPTER 4

TYPES OF CENTRIFUGATION

Centrifugation can be broken down into two major areas: preparative and analytical centrifugation. Preparative centrifugation is focused on the actual separation of cells, subcellular structures and related particles while analytical centrifugation is concerned mainly with the study of properties such as molar mass and the shape of purified macromolecules. Analytical ultracentrifugation is a versatile and powerful method used for the quantitative analysis of macromolecules in a solution. It is most commonly utilized for observing the fractionation process in real time while the sample is still being centrifuged. This is usually done by an optical detection system using ultraviolet light adsorption. Centrifugation can reach very high speeds with centrifugal forces reaching up to 1 000 000 *g*. This type of centrifugation is usually referred to as ultracentrifugation and can also be carried out using the main techniques of preparative and analytical centrifugation. Modern preparative ultracentrifuges can, however, be adapted to combine and meet the requirements of both techniques. Therefore, preparative ultracentrifugation is usually concerned with the fractionation of samples containing tissue homogenates while analytical ultracentrifugation aims to provide more information regarding the physico-chemical properties of the sedimenting particles as well as the molecular interactions of macromolecules and their possible subunits (Völkl, 2010).

PREPARATIVE CENTRIFUGATION

The main purpose of prepative centrifugation is to isolate specific particles like organelles and molecules. This technique can be used to estimate sedimentation coefficients and molecular weights and allow for larger sample sizes to be processed. Many preparative ultracentrifuges have no optical read-out functions which means that separated fractions will have to be collected and subsequently analysed after the run. Preparative centrifugation techniques can be subsequently broken down further into two major types namely differential centrifugation and density gradient centrifugation. Density gradient centrifugation can further be divided into rate-zonal and isopycnic centrifugation (Figure 13).

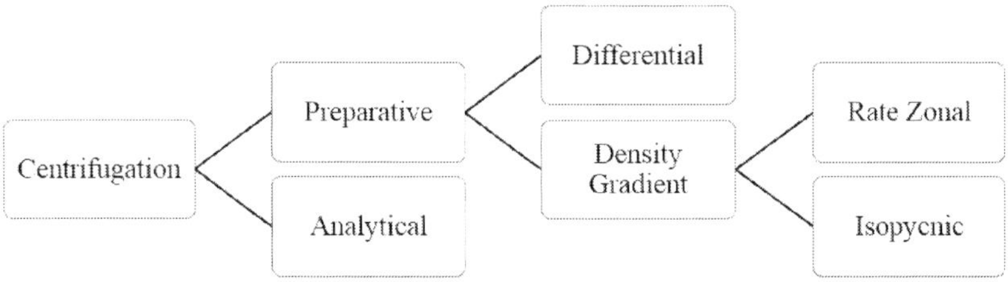

Figure 13: Outline showing the overall breakdown of preparative and analytical centrifugation techniques.

Differential Centrifugation

The simplest form of centrifugation is known as differential centrifugation, although it is more commonly known as differential pelleting. This centrifugation technique was initially developed by Stanley and Wyckoff in 1937 as an effective purification method for the isolation of the tobacco ring spot virus and tobacco mosaic virus. This method has since worked well when applied to the separation of particles with sedimentation coefficients that differ by orders of dehowever; however, it is less successful with the separation of particles that have very small differences in sedimentation rates. Differential centrifugation is commonly used for the actual harvesting of cells or producing crude extracts containing subcellular fractions of nuclei, mitochondria and even lysosomes from tissue homogenates (Figure 14). For instance, a rodent liver homogenate that contains mitochondria, nuclei, lysosomes and vesicles that is centrifuged at low speed for a brief span of time is going to pellet the bigger and more dense nuclei. These specific parts of the cell can then be subjected to further analysis. Typically, a tissue sample is first homogenized to break down the cellular membrane followed by mixing of the cell contents to form a homogenate. The homogenate then undergoes repeated centrifugation cycles yielding two fractions namely the pellet which contains the sedimented particles and the supernatant which contains the unsedimented particles together with the buffer. Pollution by various molecule sorts can be minimised by resuspension and rehashing the centrifugation steps. The supernatant can then be re-centrifuged at higher speeds to obtain a new pellet and supernatant. The pellet can also be re-centrifuged after resuspension in a small volume of a suitable solvent to wash the pellet and remove any sample solvent left behind from the pellet.

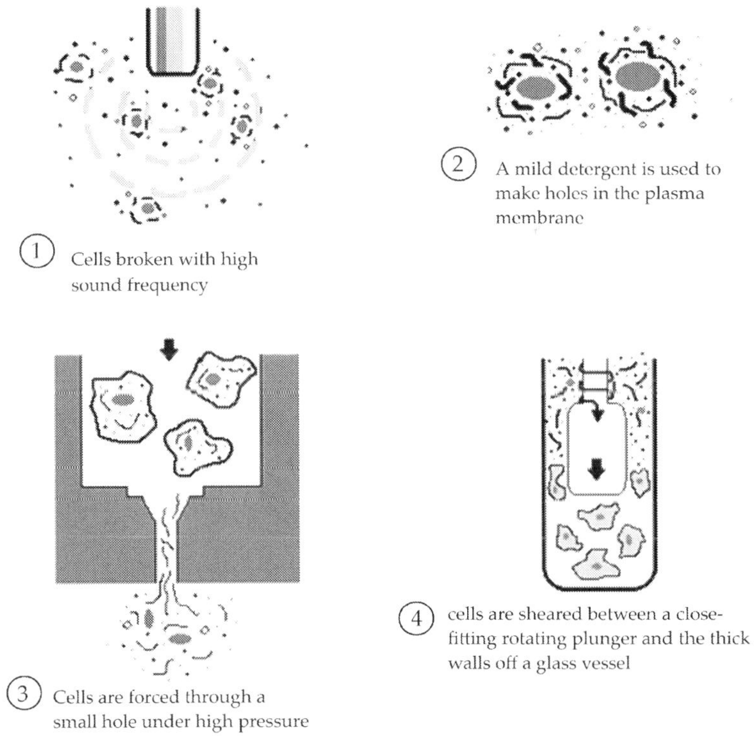

Figure 14: Common mechanical methods used to achieve homogenization and release cell contents.

For example, homogenates containing sub cellular components such as nuclei, mitochondria, lysosomes and membrane vesicles that are centrifuged at lower speeds and for shorter time periods will only pellet the larger subcellular components like nuclei. Subsequent centrifugations at higher speeds will result in the pelleting of particles of the next order of size (Table 3). By undergoing centrifugal separation, particles are thus allowed to sediment out. Particles of different sizes or densities within a suspension will separate and sediment at different rates. Larger and denser particles will sediment faster than smaller or less dense particles. For a typical cellular homogenate, centrifugation at low speeds for 10 minutes will yield a pellet containing whole unbroken cells, cell nuclei and large debris (Figure 14). This is usually referred to as a nuclear pellet. Increasing the speed to approximately 10 000 – 20 000 g for 10 minutes will allow for sedimentation of mitochondria and lysosomes. This pellet is known as the mitochondrial pellet. Further centrifugation at speeds of approximately 100 000 g for even longer periods of about 60 minutes will allow for the separation of crude microsomes and the cytool.

The sSedimentation rate can be influenced by the centrifugal force. A suspension of cells that undergos a cyclic series of increasing centrifugal force will yield a series of pellets containing cells of decreasing sedimentation rates. Initially, all particles of a homogenate are homogenously distributed throughout the solution in a centrifuge tube. Depending on the size as well as the centrifugal conditions, different components will separate out differently. During centrifugation, particles will move down the centrifuge tube at their respective sedimentation rates and start to form a pellet on the bottom of centrifuge tube. By increasing the applied centrifugal field, the material to be separated can further be divided into various fractions. Ideally centrifugation is cyclically repeated long enough to pellet most of the large particles. The resulting supernatant is then centrifuged at a higher speed to separate medium-sized particles followed by centrifugation at even higher speeds to separate smaller particles (Figure 15).

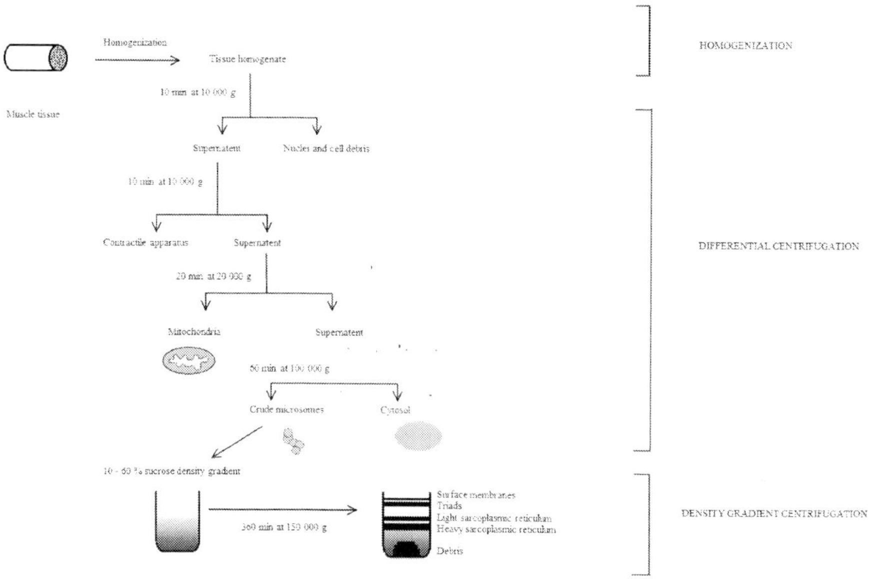

Figure 15: Fractionation methodology for membranes from skeletal muscle tissue homogenates using differential centrifugation (Wilson and Walker, 2010).

Advantages of differential centrifugation include the simplified methodology required for enrichment of cell organelles and that gradient centrifugation is greatly enhanced by using an enriched preparation of organelles rather than a crude homogenate. Despite repeated centrifugation steps, differential centrifugation, however, only leads to an enrichment of particles rather than the purification of a particular particle, thereby reducing the overall yield. One of the major disadvantages of differential centrifugation is that, due to the varying biological particles often present in homogenates, this technique is characterised by contamination with other particles and poor recovery rates. Usually in the time required for the complete sedimentation of heavier particles, some of the lighter and medium sized particles that are originally present near the bottom of the centrifuge tube will also sediment out with the heavier particles, thereby contaminating the required pellet.

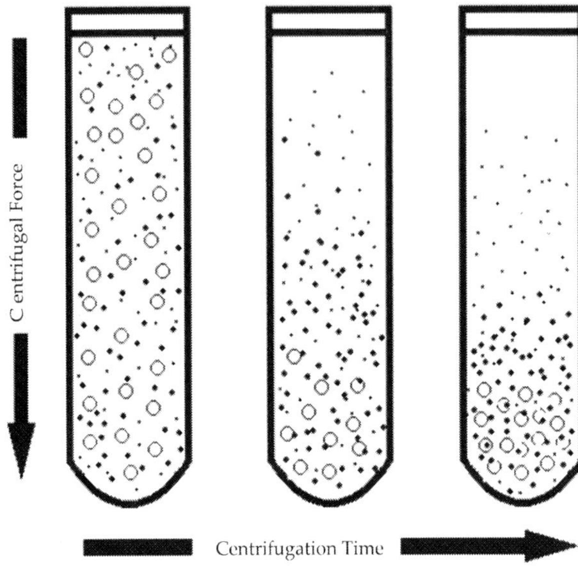

Figure 16: Differential centrifugation showing particles with different densities and sizes sedimenting at different rates. The largest and most dense particles will sediment the fastest followed by the less dense and smaller particles.

The pellet obtained after centrifugation is a mixture of all of the sediment components and is still contaminated with some unsedimented particles that were initially at the bottom of the tube. The only component of the sample which is in a near purified form is the one that sediments the slowest during the run. However, its yield is very low. This means that pure preparation of a pellet containing the heaviest particles cannot be obtained through a single centrifugation run. Each pellet, once obtained, should be resuspended separately and further separated by subsequent equilibrium density gradient centrifugation steps. The purity can however be iy repeated additional re-suspension and re-center of timesprocesses, using similar centrifugation conditions. Despite this disadvantage, differential centrifugation is one of the most commonly employed methods for the isolation of subcellular particles from homogenized tissues (Voelkl, 2010). Run times for pelleting can be shortened by using half-filled thick-walled polycarbonate tubes that carry a smaller volume. This then means that the particles will have to travel a shorter path lench in turn means less distance for the particles to travel re and hence a shortened run time.

Criteria for Successful Differential Centrifugation

- Before differential centrifugation is carried out, the tissue sample should be adequately lysed using an appropriate homogenization protocol.
- Depending on which particles are required from the separations, numerous centrifugation runs can be carried out.
- The pellet should be removed after each run while the supernatant is recentrifuged at a faster speed allowing for the next heaviest component to separated.

Table 3: Classes of centrifugation and related properties

	PROPERTIES		PELLETING ABILITIES							
	Maximum RPM $\times 10^3$	Maximum RCF $\times 10^3$	Bacteria	Animal / Plant cells	Nuclei	Precipitates	Mem-brane fractions	Ribosomes / Polysomes	Macromol-ecules	Viruses
Low-speed	10	7	Y	Y	Y	S	S	□	□	□
High-speed	28	100	Y	Y	Y	M	S	□	□	M
Ultra / micro-ultra	100 / 150	800 /900	Y	Y	Y	Y	Y	Y	Y	Y

Y = yes; S = some; M = most and indicates that it can be done but these centrifuges are not necessarily used for this purpose (Majekodunmi, 2015)

DENSITY GRADIENT CENTRIFUGATION

Density gradient centrifugation is one of the preferred methods to purify subcellular organelles and macromolecules of similar size but differing densities. This technique was first used with swinging bucket rotors for separating or purifying viruses, subcellular particles and nucleic acids (Brakke, 1958). Numerous studies have indicated the suitability of density gradient methods for the study of viruses and nucleic acids (Brakke 1960; Parish, 1972; Schumaker, 1967; Vinograd 1963). DNA and RNA have been separated using density gradient centrifugation followed by the estimation of their size by determining their sedimentation coefficients.

Studies have also indicated that cells and subcellular fractions like erythrocytes, lymphocytes, bacteria, and mitochondria have been isolated and purified using density gradient centrifugation (Skyes 1971; Birnie, 1972). Gradients are generated by placing media of varying concentrations layer after layer in a tube, in succession with the heaviest layer at the bottom to the lightest layer at the top. The cell fraction which contains a mixture of particles to be separated is then placed on top of the density gradient layers. There are two methods of density gradient centrifugation namely, rate zonal centrifugation and isopycnic centrifugation also known as iso-density or equal density centrifugation. Both methods can be used when the quantitative separation of all the components in a mixture is required. These methods are also used for the determination of buoyant densities whereby particles will be separated based on their density being similar to a particular layer in the density gradient thereby causing them to float or rise to the top of this particular layer (Table 4). In addition, the estimation of their sedimentation coefficients can also be determined.

Table 4: Densities of macromolecules in sucrose solution (Griffith, 2010)

MACROMOLECULES	DENSITY (g/cm^3)
Golgi apparatus	1.06 - 1.10
Plasma membranes	1.16
Smooth endoplasmic reticulum	1.16
Intact oncogenic viruses	1.16 - 1.18
Mitochondria	1.19
Lysosomes	1.21
Peroxisomes	1.23
Soluble proteins	1.30
Plant viruses	1.30 - 1.45
Rhino- and enteroviruses	1.30 - 1.45
Nucleic acids, ribosomes	1.60 - 1.75
Glycogen	1.70

Density gradient centrifugation, while somewhat more complicated than pelleting or differential centrifugation,ehas some advantages. The density gradient method allows for the complete separation of several or all components in a sample mixture. It also allows for analytical measurements to be obtained. Subsequent changes in the measured parameters are then used to study the effects of various chemical, physical, or biological treatments on the sample material from which it was isolated. There are a range of density gradient media available that are commonly used for separation, each with their own respective benefits for the particles they are intended to separate. Caesium chloride is widely used for the separation of DNA and the isolation of plasmids, nucleoproteins and viruses. Sodium bromide and sodium iodide are commonly used for lipoproteins and DNA or RNA molecules. Percoll, Ficoll, Dextran Metrizamide and Nycodenz are some of the commercially available gradient materials used for separation of whole cells and subcellular particles. Regardless of the density gradient medium chosen, if separation of all the components within a tissue homogenate is required then the maximum density of the gradient medium must exceed the density of the most dense particles present within the homogenate.

Production of density gradients can be done through the use of two main techniques. The first is a mechanical hand-held method while the second is done through the use of an automated gradient maker. Generally commercially available gradient makers are commonly used to establish density gradiener; however, if these automated gradient makers are not available, then the manual pouring of a stepwise gradient can be done with the help of a pipette. In addition the rigidity of the specific parameters required for separation will determine which method should be used for producing a particular density gradient. For example, if the formation of a stable continuous gradient is required then a commercially available gradient maker must be used. After pouring, gradients are usually kept in a cold room to ensure temperature equilibrium is maintained. They should be moved extremely slowly in special holders so as to avoid mixing of the different gradient layers.

Rate Zonal Centrifugation

Rate zonal centrifugation is commonly used to separate particles based on their sedimentation rates and is usually carried out with a sucrose gradient. As with differential centrifugation, the problem of contamination from other particles in the solution being centrifuged poses a major problem, especially when purified preparations are required for further downstream processing. This problem of cross-contamination from particles of different sedimentation rates can be avoided using rate zonal centrifugation in which a sample is layered as a narrow zone on top of the density gradient. The density gradient allows for the removal of particular contaminants due to the medium varying in density along the tube instead of being homogenous as in differential pelleting. In this way the faster sedimenting particles are not contaminated by the slower particles and even slight differences in a particles physicochemical properties like size, shape and sedimentation can be exploited for their separation and resolution (Voelkl, 2010).

With rate zonal centrifugation, particles that vary by about 20 to 30% in their sedimentation coefficients can be separated. This method utilizes a density gradient solution that contains a layer of sample solution which fills about 5% of the centrifuge tube capacity. In this technique, the main function of the gradient is to stabilize the liquid column within the centrifuge tube as well as to help with visualization of particles of interest. Particles are then separated by size based on differences in their sedimentation coefficient (Figure 17). The cross-contamination of particles with different sedimentation rates forms one of the biggest disadvantages with this method and is usually overcome by layering the sample as a narrow zone on top of a density gradient so that the faster sedimenting particles will not be contaminated by the slower sedimenting particles. However, this narrow zone also means that a lower volume of sample usually around 10% will be accommodated for single separations.

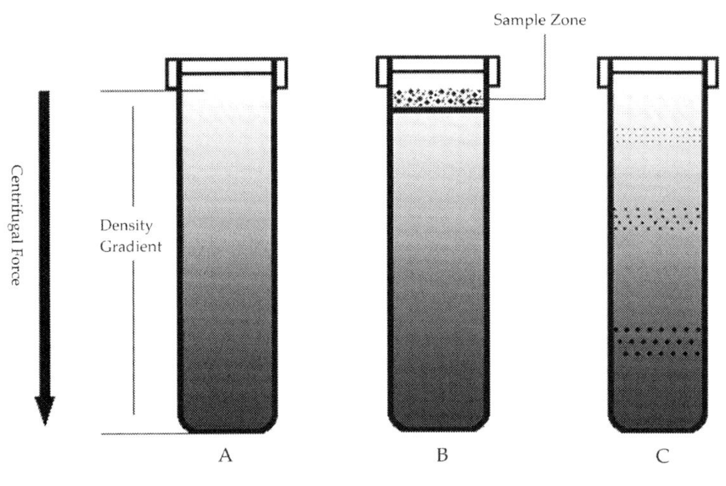

Figure 17: Rate zonal centrifugation showing a density gradient solution with a sample zone layered on top.

Initially all particles are uniformly distributed within the sample zone (A). After a centrifugal force is applied, particles move at different rates, depending on their mass, and are separated out into zones depending on their sedimentation rates (C). Initially the largest particles will sediment first with further centrifugation steps resulting in sedimentation of intermediate-sized and smaller particles in succession. A: Centrifuge tube filled with density gradient; B: Sample applied to the top of the gradient; C: Under centrifugal force particles move at different rates and separate out depending on their mass.

A swinging bucket rotor is generally used for rate zonal centrifugation separations. As the faster sedimenting particles move further ahead and down through the density medium, zones containing particles of similar size will form, thereby leading to separation. All particles will eventually form a pellet and separate out if centrifuged long enough because the density of the particles is usually greater than the density of the gradient medium. This method relies on differences in size, shape and density of the particles as well as the density and viscosity of the medium together with the applied centrifugal field. Unfortunately this method is not efficient for subcellular organelles which have different densities but similar sizes. It does however work well for the separation of enzymes, ribosomal subunits and RNA-DNA hybrids.

Criteria for Successful Rate-Zonal Centrifugation

In order to obtain successful rate zonal centrifugation separations, the following criteria should be met:

1. The density of the sample solution must be less than that of the lowest density portion of the gradient.
2. The density of the sample particle must be greater than that of the highest density portion of the gradient.
3. The centrifuge tube, and therefore the path length of the gradient, must be sufficient for the separation to occur.
4. Centrifugation should run for the appropriate amount of time to allow for adequate separations. If runs are too long then all the particles may pellet at the bottom of the tube while if the run is too short then no particles will sediment out.

Isopycnic Centrifugation

Isopycnic centrifugation is very useful for isolating membranes and organelles that have diverse sizes but similar functions, compositions and densities. This technique is also known as buoyant or equilibrium centrifugation and utilizes a density gradient solution to separate particles solely based on differences in their density. The methods used in this technique are a combination of sedimentation and flotation. This technique involves layering the sample on top of a density gradient that contains gradient concentrations that cover the whole range of particle densities present. In addition, the density of the gradient medium must be greater than the density of the particles being separated. Thie therefore separates particles into zones solely on the basis of their different densities,ndependently of time. During centrifugation, particles of a specific density will continue to move and sediment until they reach the point where their density is equal to that of the gradient medium, irrespective of their starting position. This is then referred to as the equilibrium or isopycnic position. At this point, no further sedimentation will occur even if the centrifugation run is extended for longer time periods. This is because the particles are already floating on or above a density medium that has a density greater than that of their own. It is therefore said that the particles are separated out according to their buoyancy (Figure 18). However, if the maximum density of the gradient always exceeds the density of the densest particles in solution, then the respective particles will never sediment to the bottom of the centrifuge tube regardless of centrifugation time.

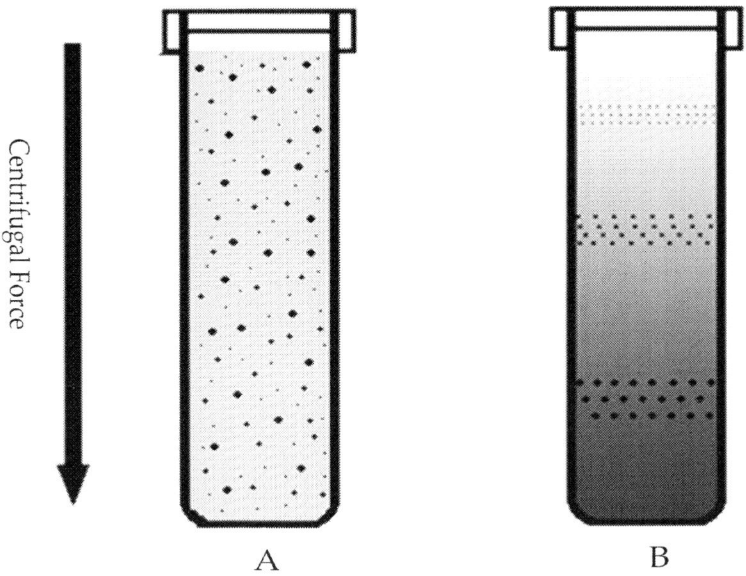

Figure 18: Isopycnic gradient separation showing particles in a sample separated to their buoyant positions. (A) Uniform mixture of sample and gradient (B) under a centrifugal force, the gradient redistributes and particles will band at their isopycnic positions.

One of the advantages associated with this technique is that when not all the present components in a solution are needed, a gradient range can be selected wherein unwanted materials will sediment at the bottom of the centrifuge tube and required particles will float to their respective isopycnic positions. This is therefore said to be a combination of both rate-zonal and isopycnic centrifugation. One of the major disadvantages of this technique is that great variations can occur, since the density of biological particles is extremely sensitive to the osmotic pressure of the gradient. This means that isopycnic separation will vary significantly depending on the type of actual gradient medium used. With this technique, particle size will only affect the rate at which particles will move until their density is the same as the surrounding gradient medium. Sub cellular organelles and particles that vary with 0.01 to 0.02 g/cm^3 in their density can be easily separated usingmethod whilstwhile soluble proteins which have similar densities cannot be separated using this method. Isopycnic centrifugation may be carried out using continuous or discontinudientshowever; however, there is no specific requirement for where the sample should be loaded relative to the gradient solution. The sample does not necessarily have to be loaded on the top. Floatation separation techniques can also be used wherein samples are loaded at the bottom of the gradient or mixed throughout the gradient. In this case the particles are said to be in a medium of density greater than that of their own density. During centrifugation particles will float up through the gradient rather than sediment through it to the bottom.This technique is commonly used for the separation of different types of lipoproteins as well as for the sub fractionation of different types of membranes (Rickwood, 2014). The sample can alternatively be placed in the middle of a discontinuous gradient or even distributed throughout the gradient. In self forming gradient techniques, the gradient is formed during the centrifugation process by the sedimentation of the gradient medium while in preforming gradients; the density gradient solution is prepared beforehand. Furthermore while both continuous and discontinuous gradients can be used with this technique, a continuous gradient is more suited for analytical purposes. Continuous gradients are used when high purity separation is required and when particles need to be separated from other particles with a similar density. These gradients have a gradual transition from high to low density such that each type of particle will band at its own specific density following centrifugation.

Preparative techniques commonly use a discontinuous gradient, which allow particles to band at the interface between the density gradient layers. This is ideal for harvesting certain biological particles (Majekodunmi, 2015). In addition, preformed gradients are used for larger particles such as cells and organelles while, in the case of macromolecules and other smaller particles self-forming gradients are used. Samples that require low purity separations will most commonly also use discontinuous gradients. These consist of a low density solution layered on top of a high density solution with a sharp difference in density between them. The interface is where all the particles with a specific density between the high and low solution concentrates. While discontinuous gradients are easier to prepare and fractionate, continuous gradients are cleaner and more accurate. A range of different substances can be used for creating density gradients (Table 5). These include sugars, salts, polyethylene glycol and synthetic polymers.

Table 5: Application of density gradient media for isopycnic separations

GRADIENT MEDIA	CELLS	VIRUSES	ORGAN-ELLES	NUCLEOPROTEINS	MACROMOLECULES
Sugars (eg.: sucrose)	+	+ + +	+ + +	+	-
Polysaccharides (eg.; Ficoll)	+ +	+ +	+ +	-	-
Colloidal Silica (eg.: Percoll)	+ + +	+	+ + +	-	-
Iodinated Media (eg.: Nycodenz)	+ + + +	+ +	+ + + +	+ + +	+
Alkali Metal Salts (eg.: CsCl)	-	+ +	-	+ +	+ + + +

+ + + + excellent, + + + good, + + good for some applications, + limited use, - unsatisfactory

The density of particles is highly dependent on the gradient medium used. For example, DNA has a density of about 1.7 g/ml in CsCl gradients while in Nycodenz gradients DNA bands at about 1.11 g/ml. Sucrose gradients are most commonly used for subcellular fractions such as mitochondria while ionic gradient media like CsCl for example, having been used for nucleic acids. The choice of rotors for isopycnic centrifugation is important too and depends on the nature of the sample. For larger particles such as organelles it is best to use a swinging bucket rotor while in the case of macromolecules better results in terms of gradient capacity and resolution should be obtained using a fixed-angle rotor.

Criteria for Successful Isopycnic Centrifugation

In order to obtain successful isopycnic separations, the following criteria should be met:

- The density of particles in the sample must be within the minimum and maximum levels of of the gradient densities.
- The centrifugation run time must be sufficient for the particles to band at their respective isopycnic points.
- Any gradient length is acceptable as long as it follows point 1 and 2.

Comparison between Rate Zonal Centrifugation and Isopycnic Centrifugation

While density gradient centrifugation is preferred over differential centrifugation because it allows for the purification of subcellular organelles and macromolecules, the two techniques of density gradient centrifugation each have their own advantages and disadvantages making one more suitable than the other depending on subsequent downstream processes (Table 6). Rate zonal separations take advantage of particle size and mass for sedimentation purposes while isopycnic separation on the other hand is used to separate particles differing in density. in isopycnic separations, size only affects the rate and time of migration (Figure. Centrifugation is performed at much lower speeds and for shorter periods of time during rate zonal centrifugation resulting in particles traveling through a steep gradient while isopycnic centrifugation sees particles traveling though more of a flat gradient. In rate zonal centrifugation, since the maximum density of the gradient does not exceed that of the most dense particle, it may result in a pellet forming after centrifugation, while isopycnic separation is based on buoyant densities of the particles so no pellet will form.

Table 6: Comparison of properties between Moving zone centrifugation and isopycnic centrifugation

	Moving zone centrifugation	**Isopycnic centrifugation**
Centrifugation speed	Lower speeds	Higher speeds
Sample	Not completely sedimented	Complete sedimentation to density equilibrium
Run times	Stop at specific time	Can have longer run times
Centrifugation principles	Depends on sedimentation rate	Depends on sedimentation equilibrium
Separation properties	Similar density, different molecular weight	Similar molecular weight, different density
Examples of commonly separated particles	Nucleic acids, cell organelles	Proteins

Common applications of rate zonal centrifugation include the separation of cellular organelles and proteins, such as antibodies which all have similar densities but differ in their masses and molecular weights, while isopycnic separations are used for isolating organelles and membranes that have very different sizes but similar functionality and composition and thus similar densities (Figure 19).

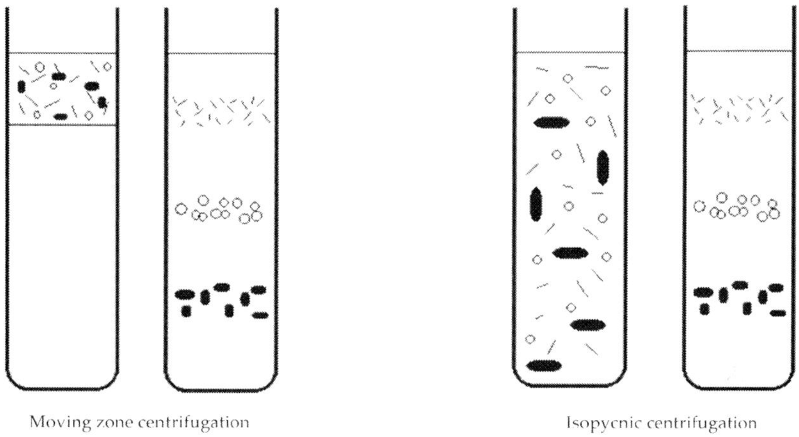

Figure 19: Comparison between rate or moving zone centrifugation and isopycnic centrifugation. In moving zone centrifugation the sample to be separated is loaded on a narrow zone at the top of the gradient while in isopycnic centrifugation the sample and gradient.

ANALYTICAL CENTRIFUGATION

While the main purpose of prepative centrifugation is to isolate specific particles, analytical centrifugation involves measuring the physical properties of the sedimenting particles. The first ultracentrifuge to be developed was an analytical ultracentrifuge wherein the sample to be separated was placed in a chamber with transparent side making it possible to view and analyse the sedimentation process while centrifugation was taking place. In this way, the sedimentation rate of a particle could be measured in real time. Today many analytical centrifuges contain light adsorption and a range of technical optics present which provide the quantitative analyses of macromolecules allowing researchers to obtain a great deal of information regarding their shape, size and compostion (Rickwood, 2014). Analytical centrifugation usually utilises a homogenous or linear density gradient solution for sedimentation such as a 5 to 20 % (w/w) or 10 to 40 % (w/w) sucrose solution as it is close to being isokinetic ensuring that particles will move at the same speed throughout the gradient. In addition, it also allows for stabilization of the liquid column and more materials to be loaded on top of the gradient solution as a stable layer. Analytical centrifugation is generally used when regular methods like gel electrophoresis fail to yield the required information. It is especially useful for studying the molecular interactions of macromolecules like polysaccharides and, while it can be expensive to purchase this piece of equipment, the type of information provided is both accurate and reliable ensuring that it is a beneficial investment.

CENTRIFUGAL ELUTRIATION

While classical methods of cell isolation are based on differential and isopycnic centrifugations, these techniques, despite their efficiey,do still present some disadvantages. These depend on the chemical nature of the products used to form a density gradient as well as other factors like osmotic pressure, viscosity or concentration (Rickwood, 1984). In addition, these methods also require several stages of centrifugation, decanting and collection. Elutriation is known as a gentle process for separating particles based on their shape, size and density using a stream of gas or liquid flowing in a direction that is usually opposite to the direction of sedimentation. This technique forms one such alternative that is both efficient, rapid and which produces purified cells in large quantities. This technique was initially developed by Lindahl (1948) and was later further developed by Beckman Instruments under the name of centrifugal elutriation (McEwen et al. 1968). In this process, the particles of interest are placed into a specially designed centrifuge chamber and subjected to a centrifugal field (Figure 20). The cells are then sequentially washed out of the rotor based on their size, using a buffer stream that flows in the direction opposite the centrifugal field. By balancing the centrifugal force against the opposing buffer flow, particles are selectively removed from the mixture.

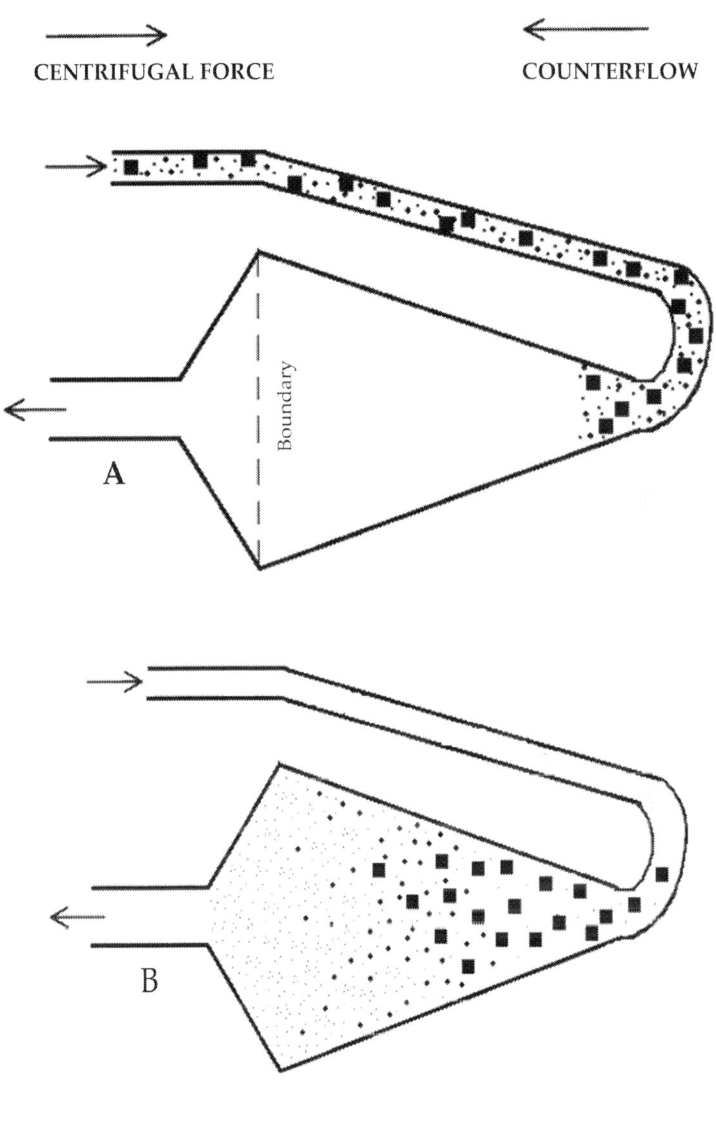
CENTRIFUGAL FORCE
COUNTERFLOW
Boundary
A
B

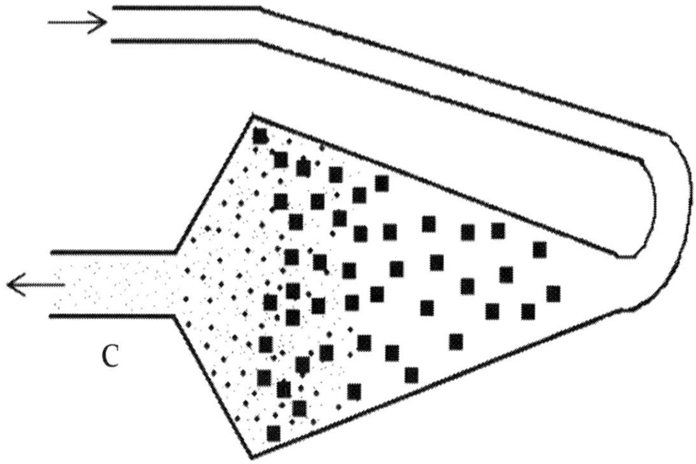

Figure 20: Illustration of a simple elutriation procedure. Particles of interest are A) suspended in an appropriate medium and then subjected to a centrifugal fel-dAs this happens the particles will enter the centrifuge chamber B) Sedimentation of particles occurs but is balanced by counter flow C) The flow is increased and slow sedimenting particles are washed or elutriated out from the chamber based on their ize. (Bachère et al, 1988).

The main principle surrounding this technique is velocity sedimentation in which cells are separated according to their rate of sedimentation. Separation is proportional to the size of the cells and to the difference between the densities of the cells in the medium. Cells are exposed to a centrifugal force while suspended in a selected medium which flows continuously in a centripetal direction. In the elutriation chamber, each cell tends to migrate to a zone where its sedimentation rate is balanced by the flow rate of the fluid. As a result of the chambers design, a gradient of varying flow rates is produced from one end to the other which inevitably means that cells with a wide range of sedimentation rates can be held in suspension and subsequently washed (Bachère *et al.*, 1988).

This technique is used for the separation and purification of a large variety of cells of different sized and from different tissues and species including the separation and purification of vertebrate cells, cells from the brain and liver and tumour cells. During separation, cells are suspended in a buffer solution. A density gradient is not required for this technique and any medium that is compatible with the particles to be separated can be used. Generally, pelleting does not occur and fractionation of particles ranging in size from 5 to 50 μm can be achieved (McEwen et al, 1968). Any buffer and culture medium may be used as long as the medium's capacity can maintain the physiological integrity of the cells and prevent cell aggregation. As a result of this technique, living cells can be separated without chemical damage and can be done aseptically. This results in quick separation and high cell concentrations yields together with a good recovery yield allowing separated cells to then be used for further experimentation. In addition, since cell size is often correlated with different stages of the cell cycle, this method can also be used to separate cells at different stages of the cell cycle.

Centrifugal elutriation (CE) is currently a broadly utilized preparative cell partitioning method. Keeping in mind that the end goal is to streamline the division of cells that show little contrasts in sedimentation speed; a few conditions that may impact the determination limit were further studied. These conditions include rotor speed, counterflow, cell load, thickness and consistency of the elutriation medium.

Tests completed with human red platelets (rbc) demonstrated that aselective misfortunes of rbc from the rotor caused by the jetstream, could be anticipated if the partitions were carried out at high rotor speeds, as anticipated by the hypothesis. What's more, high cell loads (5×10^8 rbc) brought about preferable partitions over low cell loads (5×10^7 rbc). Human monocytes were isolated into subpopulations that contrasted just about 0.003 g/mL in thickness, however have basically a similar size. The detachment was checked out either by expanding the thickness or consistency of the elutriation medium or by diminishing the rotor speed. In all cases comparable outcomes were acquired. These outcomes demonstrated that under ideal conditions centrifugal elutriation can be connected for the partition of cells that contrast just marginally in sedimentation speed.

A recent study researched whether centrifugal elutriation could be streamlined such that it could be connected to the cleaning of cells that contrasted just marginally in their sedimentation speed. Hypothetical results revealed that the state of the partition chamber can in fact impact the determination. Here it is shown that if the different elements that influence the determination are dealt with ideally, subpopulations of human red platelets and of human fringe blood monocytes can be disconnected. Centrifugal elutriation (CE) was done with a Beckman JE-6 elutriator rotor outfitted with the standard detachment chamber in a J-21C rotator. Cells were elutriated at a steady stream rate of 18 mL/min by a stepwise reduction in rotor speed or a stepwise increment in the density of the medium. The divisions were gathered in volumes of 150 mL. To shift the thickness of the centrifugal elutriation medium, the elutriation framework was adjusted. Quickly, two vast level stainless steel supplies R1 and R2 (figure below) containing centrifugal elutriation and centrifugal elutriation-P medium, individually, were set at a tallness of 180 cm and were loaded with 2 L of medium. The thickness of the liquid layer in these stores is around 3 cm.

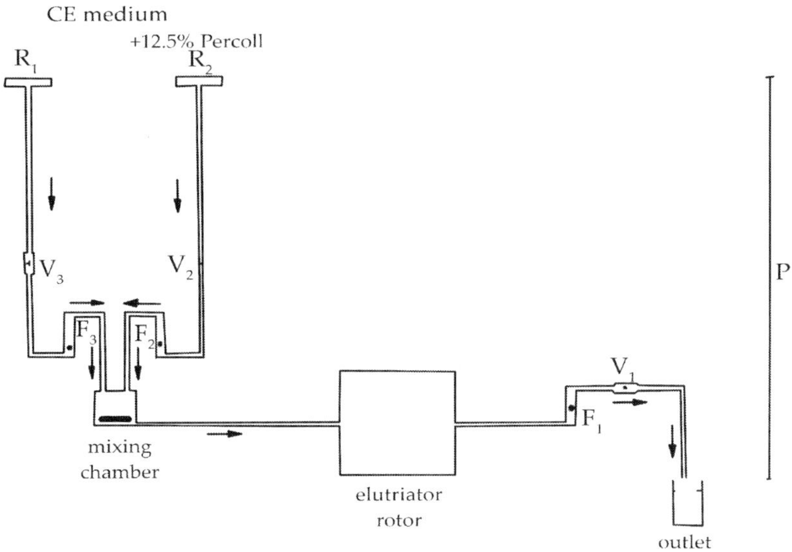

Figure 21: Schematic representation of elutriation system.

Since elutriation is conveyed out at 18 mL/min and the weight is 180 cm, the stream rate will diminish. Variety of the density of the medium was proficient by blending centrifugal elutriation medium (thickness = dl) with centrifugal elutriation-P medium (thickness = d 2), as portrayed in the legend to the following figure. The thickness of the medium, dm, is given by:

dm= (F3dl + F2d2)/Fl = dl + (d2 - dl)(F2/F1)

in which F = stream rate.

ULTRACENTRIFUGATION

Ultracentrifugation has advanced the study of subcellular structures by allowing for detailed biochemical analyses to be conducted through high speed centrifugation, allowing rotors to achieve high speeds of around 5000 rpm (Figure 25). Svedberg designed and developed the first analytical ultracentrifuge whereby the sample was centrifuged in a cell with transparent sides. This allowed for viewing and analysis of the sedimentation process during centrifugation. Ultracentrifuges generally allow for the sedimentation velocity experiments and sedimentation equilibrium experiments to be carried out. The main aim of sedimentation velocity experiments is to determine the total time taken for sedimentation to occur as well as to provide additional information on the shape, size, distribution and molar mass of the particles of interest. These kinds of experiments can also be used to study reversible chemical equilibria between macromolecular species. This is usually done through monitoring the number and molar mass of macromolecular complexes present. Sedimentation equilibrium experiments on the other hand focus on the final steady state of an experimental run. Sedimentation velocity experiments are usually performed at high speeds. The main purpose of this is to overcome the effect of diffusion. To carry out this method, an initial uniform solution is placed in a cell and a high angular velocity is applied to cause rapid sedimentation of the solutes towards the cell bottom. As a result, there is a depletion of solutes near the meniscus which forms a sharp boundary between the depleted region and the solutes which are sedimenting. From this, the sedimentation coefficient can be determined. Sedimentation equilibrium experiments have a lower rotor speed than sedimentation velocity experiments. In this method solute particles do not pellet at the bottom but instead the process of diffusion opposes the process of sedimentation until after a period of time when the two opposing forces reach equilibrium.

There are two main techniques of ultracentrifugation, namely preparative ultracentrifugation and analytical ultracentrifugation (Table 7). Preparative ultracentrifugation is aimed at isolating and purifying specific kinds of particles such as subcellular organelles while analytical ultracentrifugation is aimed at studying molecular interactions between macromolecules and analysing properties such as the molecular weights of sedimenting particles.

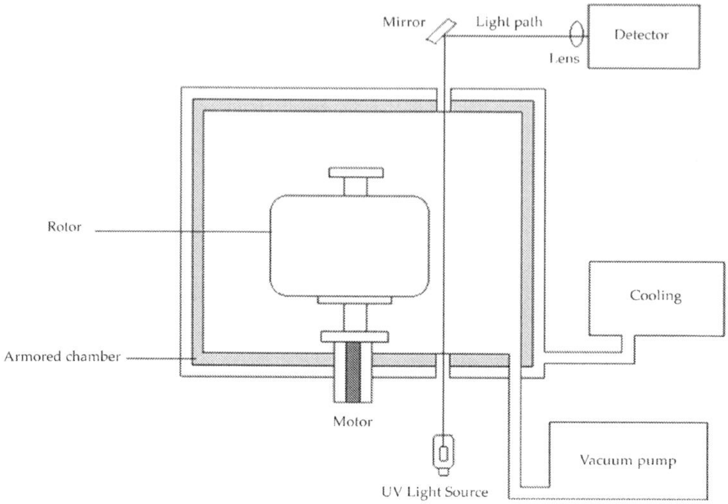

Figure 25: Overview of a generalized ultracentrifuge.

Preparative Ultracentrifugation

Modern ultracentrifuges are adapted to meet the requirements of both preparative and analytical centrifugation. Instruments are equipped with a scanning optical system needed for analytical studies in addition to the basic devices common to an updated preparative ultracentrifuge. One of the main objectives of preparative ultracentrifugation is the isolation of specific particles such as subcellular organelles which can be reused. Preparative ultracentrifugation can further be divided into two major categories – differential and density gradient ultracentrifugation. Differential ultracentrifugation is a common technique used in microbiology and cytology for the separation of certain organelles from whole cells for further analysis of specific cellular parts. Density gradient ultracentrifugation, just like low speed density gradient centrifugation, is based on differences in density and can further be broken down to rate zonal or isopycnic ultracentrifugation. A common example of the use of density gradient ultracentrifugation is in the isolation of lipoproteins. A range of studies show that isolation can occur via floatation of lipoproteins at different salt densities. These salts include NaCl, KBr and NaBr (Straatmann, 2003).

Analytical Ultracentrifugation

Analytical ultracentrifugation was a technique used in the improvement of organic chemistry and sub-atomic science. It was devised in the 1920s and over the more recent eight decades, an abundance of essential learning has been collected with respect to the specialized execution of scientific ultracentrifuges, the hypothetical establishment of ultracentrifugation in thermodynamics and physical science of macromolecules , the numerical investigation of ultracentrifugation tests, and the use of analytical ultracentrifugation to the investigation of proteins. While the use of this technique began to decline in the 1970s and 1980s, computational and instrumental advances in the 1990s significantly encouraged the investigation of protein studies which in turn resulted in a surge of enthusiasm for this technique again.

The main aim of analytical ultracentrifugation is to measure the physical properties of sedimenting particles. These properties include sedimentation coefficients, molecular weights and thermodynamic properties of macromolecules within a solution. It also allows for the determination of a number of components and species, detection of impurities, determination of the molar mass of each species and determination of the shape and charge of molecules based on their sedimentation behaviour. Generally a range of samples can be analysed using this technique. These include viruses, cellular organelles, proteins, polysaccharides, nucleic acids, ligands and even gases. A typical analytical ultracentrifuge can generate a centrifugal force of up to 250 000 g. A sample subjected to ultracentrifugation can be monitored in real time using an attached optical detection system. This system usually uses ultraviolet light adsorption and an optical refractive index system for analyses. As the rotor turns, the image of the cell or protein is projected via the optical system on to a film or computer. The concentration of the solution at various points within the cell is then determined by adsorption of a light of the appropriate wavelength. This is usually done by measuring the degree of darkening on a photographic film or by the deflection of the recorder of the scanning system which is fed into a computer. This allows for direct real time observation of the separation of a sample.

Table 9: Major differences between analytical and preparative ultracentrifugation

ANALYTICAL	PREPARATIVE
Uses small sample sizes that are less than 1 ml	Uses larger sample sizes
Contains a built in optical system to analyse the progress of sedimentation of molecules during centrifugation	Does not contain an optical system. Sedimented fractions must be collected and analysed after the run
Relatively pure samples should be used	Less pure samples can be used
Can be used to accurately determine sedimentation coefficients and molecular weights of molecules	Can be used to estimate the sedimentation coefficients and molecular weights of molecules

The amount of material required for analytical ultracentrifugation is commonly in the request of a couple of hundred micrograms. In addition, because of the fixation gradients set up amid centrifugation, a 10 to 1000 fold fixation run is normally monitored, and a size-scope of three requests of magnitude in molar mass can be canvassed in a solitary test. Communicating parts under examination may have sizes going from peptides to extensive multi-protein structures. Today, the numerical points of interest associated with such techniques are mostly embodied in the programming of machines.

CHAPTER 5

GRADIENT MEDIUMS

SORTS OF GRADIENTS

One of the first density gradient centrifugation techniques was developed in the 1950s and used a buffered sucrose solution for the purification of cellular organelles. Since then, sucrose has quickly become the density gradient medium of choice. Following this, caesium chloride gradients were used to separate DNA of different densities. In 1958, Meselson and Stahl used a caesium chloride density gradient to aid separation of DNA components during centrifugation. Subsequently, colloidal-silica suspensions were then manufactured and in 1977, the stabilized silica colloid coated with polyvinylpyrrolidone (also known as Percoll) became available for separating cells and subcellular particles. In 1968 polysaccharides were used for the isolation of mononuclear cells from circulating blood and bone marrow which eventually led to the development of the first non-ionic iodinated density gradient medium known as metrizamide. Today there are a large selection of density gradient media currently available for use in centrifugation.

A density gradient is created by layering media of increasing density in a centrifuge tube. When a sample is layered on top of a density gradient and centrifuged, the various particles move through the gradient at different rates resulting in particle separation through the formation of bands or zones in the gradient. The denser the particles, the further the migration within the centrifuge tube. The functions of density gradients vary depending on the type of centrifugation technique chosen. In rate zonal centrifugation the main function is to aid centrifugation when separating particles while in isopycnic centrifugation the main function of density gradients is to band the particles at their buoyant densities. These functions are usually carried out by preventing the formation of convection currents, stabilising the column of liquid in a centrifuge tube, preventing the mixture of separated particles, improving resolution of the separated components as well as eliminating unnecessary mechanical vibrations and thermal gradients that may disturb the migration of particles. All density gradients of varying concentrations can be produced by techniques that fall into two major categories namely discontinuous or step gradient techniques and continuous density gradient techniques.

Discontinuous Gradients

Discontinuous gradients are generally used for preparative techniques in which particles will band at the interface between the density gradient layers. These gradients are generally used for rate zonal and isopycnic separations and are made with a range of commercial solutions like sucrose, Ficoll, Percoll, Nycodenz or OptiPrep. Discontinuous gradient techniques were first used together with swinging bucket rotors for separating or purifying viruses, subcellular organelles and nucleic acids (Brakke 1958). Two methods are generally available for the production of discontinuous gradients and these include the overlaying method and thegunderlying method.

Continuous Gradients

Continuous density gradients are gradients in which the density changes evenly from one extreme to another. These gradients are usually used during rate-zonal centrifugation and are most commonly used for the separation of proteins. The density gradually decreases down the tubr; however, the slope may be linear, convex or concave. The simplest and most common continuous gradient used for separation is a linear gradient which displays a continuous decrease in solution density from the bottom of the centrifuge tube to the top (Figure 25). Linear gradients are usually generated by either a gradient maker or through a self-generating method and are useful for the fractionation of complex samples since the gradual increase in density allows for the greatest resolution of sample components.

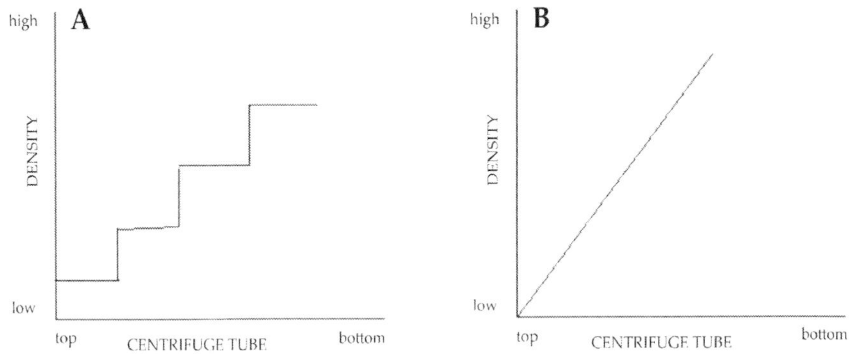

Figure 26: Comparison of A) discontinuous or step and B) continuous or linear density gradient centrifugation profiles.

The maximum density of a continuous gradient should always be greater than that of the particles in a sample. One way of improving the resolution of particles by size is to minimize the difference in relative centrifugal force between the top and bottom of the tube. This is achieved through layering the sample as a narrow band over the continuous density gradient. By allowing the particles to subsequently sediment through the density gradient, resolution and recovery is increased. Particles then move down through the gradient column at a rate that is dependent on their size and eventually will form noticeable zones (Figure 26). As the viscosity of the medium increases, the rate at which particles move is decreased. Furthermore, as the radial thickness between different particles in different zones increases, the resolution decreases further indicating that rate zonal fractionation is limited by sample size. The sample size generally forms no more than 10% of the total gradient volume. Continuous gradients are the preferred option for rate zonal separations as discontinuous density gradients may result in irregular particle separations should sudden changes in viscosity and density occur.

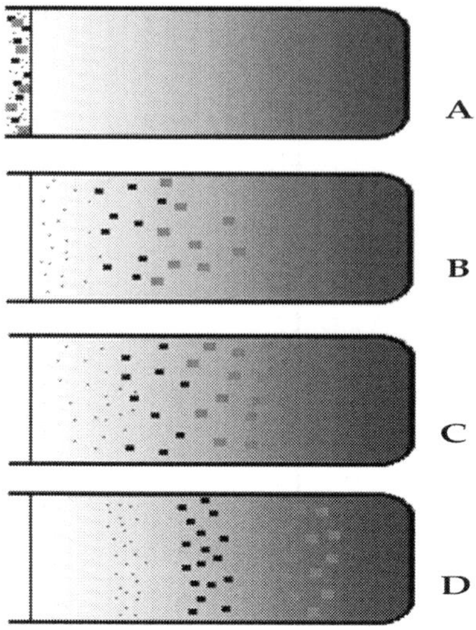

Figure 27: Density gradient separation of particles in a sample using a continuous gradient.

FORMATION OF GRADIENTS

The primary function of density gradient centrifugation is to separate whole cells and subcellular organelles from plant or animal tissue homogenates and for the purification of some viruses. This is either on the basis of their buoyant density or their rate of sedimentation (Rickwood, 2001). Both isopycnic and rate-zonal separations involve the use of a gradient solution to separate particles of interest. There are a range of different types of gradients and methods that can be used for separations. Continuous gradients are usually preferred for isopycnic separations since they are less likely to produce artefacts while self-forming gradients of concentrated salt solutions are usually used for the separation of macromolecules. For protein separation using rate zonal centrifugation, continuous linear gradients are usually used (Yamazaki, 2000).

Formation of Discontinuous Gradients

Discontinuous gradients are produced either by the overlaying techniqurlayingunderlying technique using an automatic gradient maker or the hand layering method.

Formation of Discontinuous Gradients using the Overlaying Technique

The overlaying technique is one of the most widely used methods for producing discontinuous gradients. It involves starting with known volumes of the densest solution which is allowed to run slowly down the side of the centrifuge tube followed by successive layering of the solution with lower densities on top using a pipette. The higher the viscosity of the medium, the easier it is to achieve a relatively slow and smooth flow of liquid from the pipette. Low density solutions, having very low viscosities are more difficult to layer in this manner.

Formation of Discontinuous Gradients userlaying Underlying Technique

Although the over layering technique is more commonly userlayingunderlying technique is generally easier and therefore recommended. In this technique, a layer of the lightest density solution is introduced into thc centrifuge tube followed by sucerlayingunderlying using a pipette that has its tip placed at the bottom of the centrifuge tube. A series of layers of increasing density is subsequently added and displaces the previous layer of less dense solution. The only important requirement is that no air bubbles are introduced which may disturb the lower density layers above. The discontinuous gradients can be used directly by layering of the sample to be separated as a narrow zone on the top layer which is usually of lowest density. The tube with density gradient and sample to separate is then subjected to centrifugation under the appropriate conditions.

Preparation of Discontinuous Gradients using the Hand Layering Technique

This technique utilises a percentage of a weight / weight (w/w) solution rather than a percentage of weight / volume solution because sample particles are separated either by density or size. The hand layering method requires equal volumes of the appropriate concentration of gradient solutions to be used, for example 10% w/w, 20% w/w, 30% w/w, and 40% w/w sucrose solution. These gradients can also be formed using other viscous materials (Griffith, 1986). A w/w solution is made by mixing a fixed weight of solute in 100 grams of solvent to give the required percentage of a w/w solution. For example, a 40% w/w solution is 40 grams of solute mixed in 60 grams of solvent. The concentrations should then be layered in the centrifuge tube using a pasture pipette that reaches to the bottom of the tube. The lightest concentration of the gradient (10% solution) should be placed in the tube first followed by the 20% solution which should be layered below the 10% layer. The 30% is then layered under the 20% concentration and finally the 40% concentration is layered below the 30% layer. Care must be taken not to disturb the interfaces between the layers by holding the tip of the pipette to the wall of the tube when removing or introducing it into the tube (Figure 28).

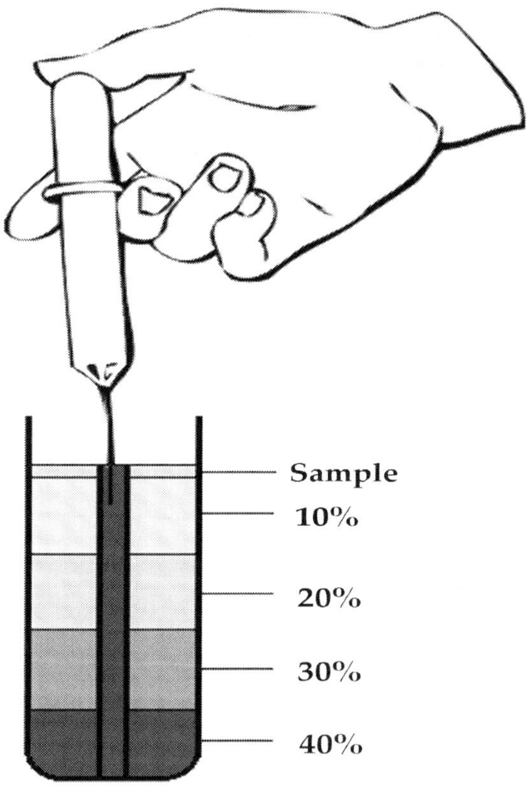

Figure 28: Preparation of discontinuous or step gradients using a simple hand layer method.

The gradient should be prepared with a cold solution and used immediately to prevent premature diffusion of the layers. The sample containing particles to be separated is first mixed with a buffer and then layered on the top gradient layer. If the sample contains particles of a known density close to that of one of the layers in the gradient, then the sample solution should be mixed with that layer of the gradient concentration prior to layering. Before applying the sample to the gradient column, the sample concentration and volume must first be determined. It is important to note that if the sample concentration on the gradient is too high, a streaming effect will result. This overloading of the gradient will cause broadening of the separated zones and loss of resolution. If the sample concentration is too low, it may be difficult to identify some of the separated zones. A fixed angle rotor is generally preferred to separate and purify particles using discontinuous gradients. A study carried out by Griffith (1986) showed that fixed angle rotors can be used to effectively separate and purify particles. After the gradient and sample is prepared and layered in the tubes, the filled tubes are capped and placed in a pre-cooled rotor. Slow acceleration to approximately 800 rpm should be used so as to prevent premature mixing of the gradient layers and the sample-gradient interface (Griffith, 2009). This slow acceleration permits the sample and layers to be oriented parallel to the axis of rotation before the rotor is rapidly accelerated to its maximum speed for the separation. During acceleration however, the gradient changes its orientation and the layers may diffuse to become linear.

Formation of Pre-Forming Gradients

Pre-forming gradients are created when the gradient is created prior to centrifugation by a gradient generator or by a hand layering method. They are prepared by either diffusion of a step gradient consisting of 3 to 4 layers or solutions or by using one of the two-chamber gradient mixers. The overall time for preparation usually takes several hours. Pre-formed gradients can also be prepared by freezing and thawing a uniform solution but the disadvantage of this technique is that it is difficult to control or predict the exact shape of the gradient. Typical density gradient solutions used in pre-forming gradients are sucrose solutions. A sample solution containing particles to be separated is layered on the density gradient solution which is prepared beforehand (Figure 29). The sample volume usually occupies 10 to 20% of the centrifuge tube capacity. The particles can then be separated by floating or sedimentation.

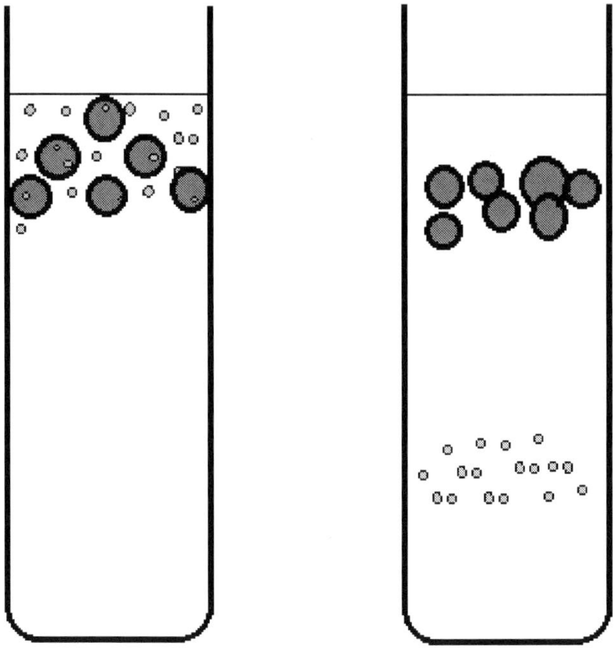

Figure 29: Separation of particles into bands using a preformed density gradient.

Formation of Self-Forming Gradients

In self forming gradients, the gradient is formed by the centrifugal field during centrifugation as the gradient medium itself starts to sediment. Under centrifugal forces the gradient material redistributes to form the required concentration and density range running from the top of the tube to the bottom. During centrifugation, the sample particles which were initially uniformly distributed together with the unseparated gradient material will either sediment or float to their isopycnic or buoyant positions (Figure 30). If the density gradient is maintained by the centrifugal force or even generated by it, then the gradient is said to be self-forming or self-generating. A commonly used density gradient solut formingself-forming gradients is caesium chloridehoer ; thehowever, the choice of density gradient used will differ depending on the type of components to be separated. For macromolecules, such as DNA, RNA and proteins, concentrated salt solutions such as caesium chloride (CsCl), potassium iodide (KI), potassium bromide (KBr), sodium iodide (NaI) and rubidium chloride (RbCl) are used. While non-ionic gradient media such as Nycodenz and Optiprep can be used to obtain better separations, these are strong salt solutions that typically disrupt the majority of biological compherefore Therefore, it is better to use sucrose gradients for organelles and membrane separations. Gradients for separating cells need to be isotonic throughout hence self-forming Percoll gradients are often used together with an osmotic balancer. However, again, Percoll is not ideal because of its colloidal nature. When using self-forming gradients it is necessary to ensure that the centrifugation conditions used optimize the shape and resolution of the gradient. When all the molecules present are at their equilibrium positions, the centrifugal force condenses the molecules into sharp zones and maintains them at that position. Once the solute begins to sediment through the solvent a concentration gradient is formed. With a sufficiently high RCF at equilibrium, the sedimentation of the solute is exactly balanced by diffusion and the gradient is said to be stable. The shape of the gradient that is formed for a particular solute thus depends on a range of factors such as sedimentation rate, path length of the rotor, time of centrifugation, speed of centrifugation and temperature. In addition, self-forming gradients will give better resolution when centrifuged in a fixed-angle or near vertical rotor.

The main advantages of using self-generated gradients are the ease of sample handling and the great reproducibility of the gradient density profile under a particular set of centrifugation parameters. The self-generating gradient technique may require long hours of centrifugation for the gradient material to form. For example, isopycnic banding of DNA takes approximately 36 to 48 hours in a self-generating CsCl gradient using swinging bucket and fixed angle aluminium rotors (Anderson, 1962).

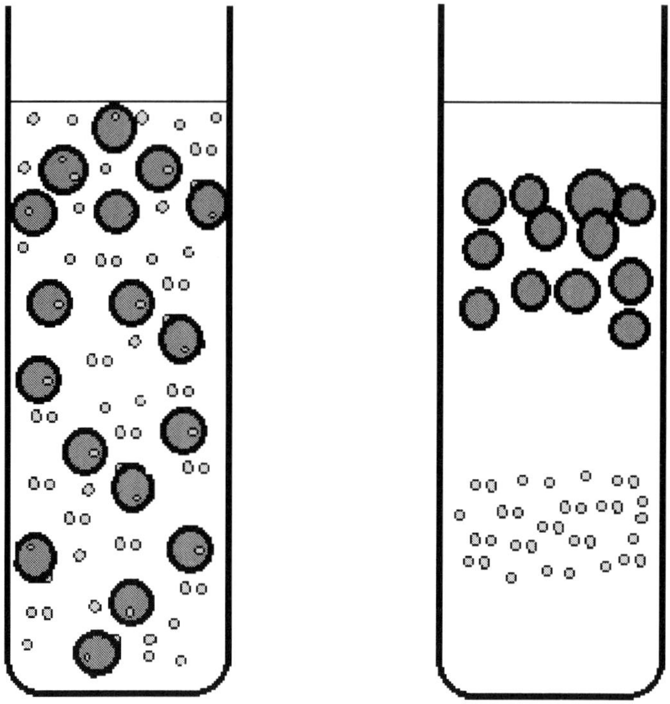

Figure 30: Formation of a self-forming density gradient solution. Prior to centrifugation, tubes contain a uniform mixture of the density gradient medium and sample with particles to be separated while after centrifugation, the particles will have sedimented at th.

Sucrose gradient Centrifugation using Discontinuous Gradients

While many substances are currently available for density gradient centrifugation, sucrose is one of the most commonly used materials for the separations of particles. Sucrose itself is a disaccharide consisting of one glucose and one fructose molecule (Figure 30). The overall structure and properties indicate that it possesses many of the characteristics of an ideal centrifugation medium, such as being relatively stable, inert to biological molecules and is relatively inexpensive to use. One of the major disadvantages however, is that sucrose solutions with a relatively high density is hypertonic and high in viscosity rendering it not useful when separating osmotically sensitive particles. This type of centrifugation is most often used to purify enveloped viruses with densities ranging between 1.1 to 1.2 g/cm³. It can also be used to purify ribosomes and exosomes. There are two methods of sucrose gradient centrifugation most commonly used. These are equilibrium centrifugation and non-equilibrium centrifugation.

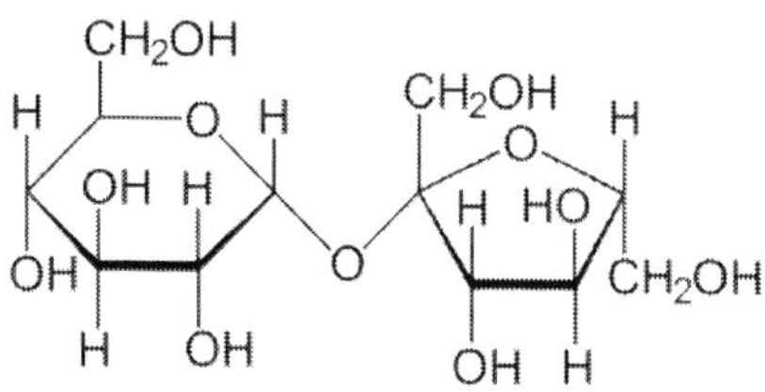

Figure 31: Structure of sucrose.

Sucrose Equilibrium Centrifugation using Discontinuous Gradients

In equilibrium centrifugation, a sucrose density gradient is created by overlaying lower concentrations of sucrose on higher concentrations in a centrifuge tube. Alternatively, a gradient mixer can be used to form the gradient layers. These devices can be used to create linear, concave, convex and exponential gradients. For example, a sucrose gradient may consist of layers ranging from 70% sucrose to 20% sucrose changing by 10% increments (Figure 31). The level of increments however, is variable depending on the sample to be purified.

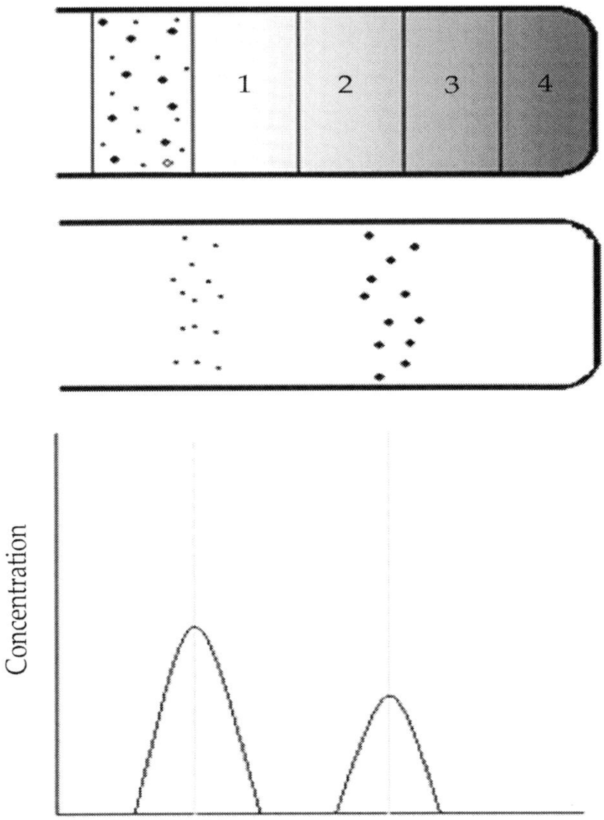

Figure 32: Rate Zonal Centrifugation using a sucrose density gradient. The sample is applied in a thin zone at the top of the centrifuge tube on a sucrose density gradient that changes in 10% increments. Density 1 < density 2 < density 3 < density 4. Under the app.

Samples containing particles of interest to beeparated areis then layered on top of the gradient and centrifuged at forces greater than 150 000 x g. Particles will subsequently travel throughout the sucrose density gradient until reaching a point in the gradient at which their density matches that of the surrounding sucrose solution. At this point, equilibrium has been reached and the fraction can then be removed and analysed further.

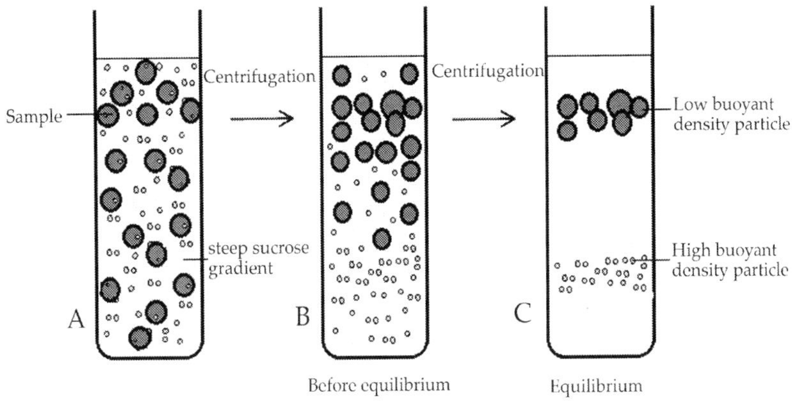

Figure 33: Separation of subcellular components using a sucrose gradient equilibrium centrifugation method.

In this technique, a steep sucrose gradient is used and cellular components are separated only on the basis of their buoyant density and not on their shape or size. The sample is dispersed throughout the sucrose density gradient (A). During centrifugation, each subcellular component will move up or down (B) until reaching a position where its density matches it surroundings. Once this happens equilibrium is achieved and no further movement will occur. A series of distinct bands will be produced with those nearest to the bottom of the tube containing the components of highest buoyant density (C).

Sucrose Non-Equilibrium Centrifugation using Discontinuous Gradients

Non-equilibrium centrifugation is very similar to equilibrium centrifugation but, in this technique, the experiment will only run until a certain point and the run is usually ended before equilibrium is reached and once banding of the required particles occurs (Figure 34). Gradients used in this technique are usually called velocity gradients. The desired particles will be at a set distance from the surface, and that band of the gradient can then be collected. After the run, collection of the desired fraction can be done by one of several methods. The easiest way is for the sucrose solution to be eluted or drained by puncturing the bottom of the tube. Only the part where the desired particles are found is eluted and thus separated. Another method includes gently suctioning out the sucrose which is done in such a way that division of the suctioned material into successive fractions occur, leading to separation. These can then be studied further to determine the distribution of the desired material, which allows you to choose the fractions containing the particles of interest.

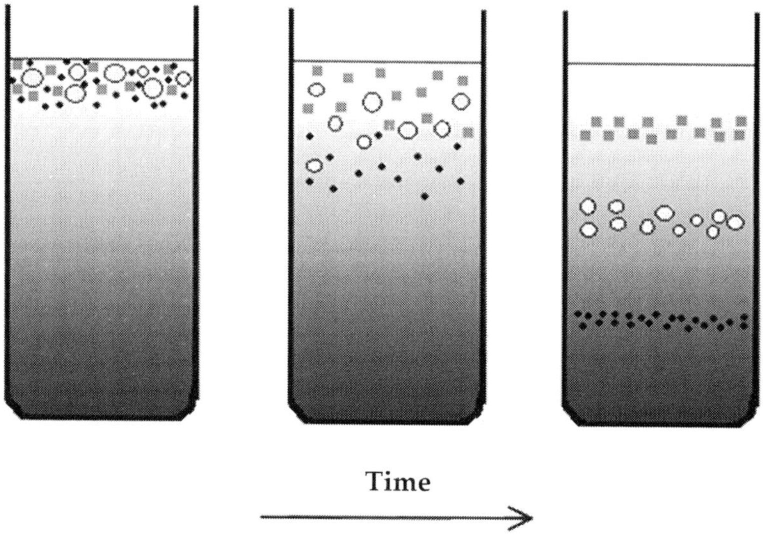

Figure 34: Separation of particles using a sucrose non-equilibrium centrifugation technique.

A mixture of three different components is initially layered onto a sucrose gradient. Application of a centrifugal force over a certain period of time will result in separation. Centrifugation will run only until the particles of interest have banded and will be stopped before equilibrium is reached.

Sucrose Cushion Centrifugation using Discontinuous Gradients

A sucrose cushion is one of the simplest forms of a discontinuous density gradient method, in which the sucrose concentration increases from the top to the bottom of the centrifuge tube in increments. It is generally used to separate specific types of particles with clearly differing sedimentation coefficients. In this technique a particle mixture is pelleted through a 20% sucrose layer and eventually comes to rest at the interface with a

70% solution (Figure 35). This allows the concentration of particles from a of interest. WhilstWhile standard centrifugation pellets the particles against the bottom of the centrifuge tube, the sucrose cushion method causes no mechanical stress and allows the collection of morphologically intact particles.

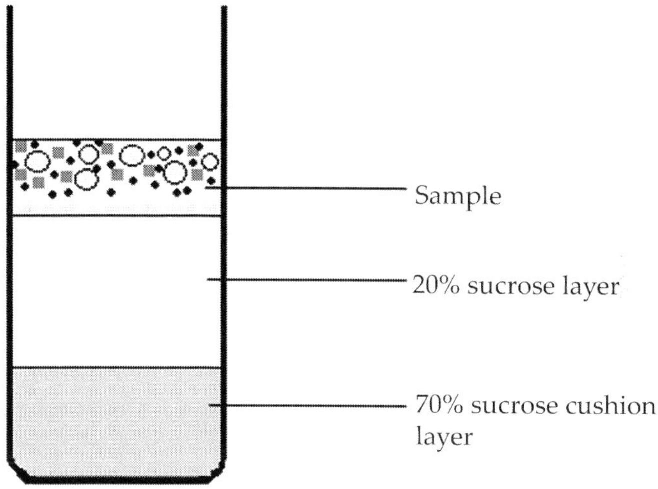

Figure 35: Centrifuge tube with a discontinuous gradient containing a 70 % sucrose cushion layer.

Formation of Continuous Gradients

Continuous gradients can be generated using a specially purchased gradient maker or by allowing discontinuous gradients to diffuse over time.

Formation of Continuous Gradients by Diffusion of Discontinuous Gradients

If the discontinuous gradient is allowed to stand for a certain period of time, the layers begin to slowly merge by diffusion to produce a continuous linear gradient. The sharp boundaries between the layers start to disappear as the solute molecules diffuse down the concentration gradient from successive denser layers to lighter layers. In this way, the difference in density between each layer will slowly even out resulting in the gradient eventually becoming linear (Figure 36). Given sufficient time the density will become completely uniform. Depending on the medium chosen, the rate of diffusion across an interface will depend on the temperature and the area and distance of the interface. Thus a linear gradient will form more rapidly at room temperature than at 4°C and if the distance between interfaces is reduced. This diffusion rate is dependent on a range of factors such as temperature, concentration range and viscosity of the gradient, thickness of each layer and relative molecular mass of the solute. Once a discontinuous gradient is formed, the sharp boundaries between the layers start to disappear as the solute molecules diffuse down the concentration gradient from each denser layer to subsequent lighter layers. Over sufficient time, the gradient will eventually become completely uniform.

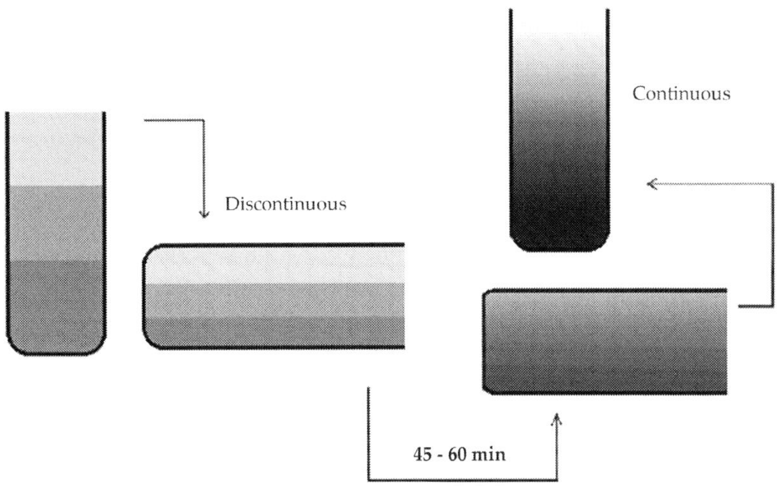

Figure 36: Formation of a continuous linear gradient from a discontiuous gradient by diffusion over an extended period of time.

Formation of Continuous Gradients using a Two-Chamber Gradient Maker

The traditional way of constructing a continuous gradient is to use a standard two-chamber gradient maker which consists of two identical chambers connected closely to their bases by a channel (Figure 36 and 37). The apparatus consists of two chambers known as the mixing chamber and the back reservoir or non-mixing chamber. The mixing chamber has an outlet directly opposite the inlet from the first channel. The device is set up with the mixing chamber resting on a magnetic stirrer and the outlet tube leading to the bottom of the centrifuge tube via a peristaltic pump. The high density solution that is needed to be separated is placed in the non-mixing chamber and the tap is opened to fill the connecting tube. An equal volume of the low density solution is placed in the mixing chamber. Two stirring bars are placed in the two chambers to ensure that the heights of the two solutions are the same. When the pump is switched on, the magnetic stirrer starts up and the connecting tap is opened. As the levels in the two chambers fall simultaneously, the speed of the stirrer is reduced to avoid generating air bubbles which may enter the gradient and disturb it. Generally there are two ways of making a continuous gradient using a gradient maker. These include A) the lower density method (Figure 36) and B) the higher density method (Figure 37). Both methods are based on the mixing of two solutions of different density before adding them to the centrifuge tube.

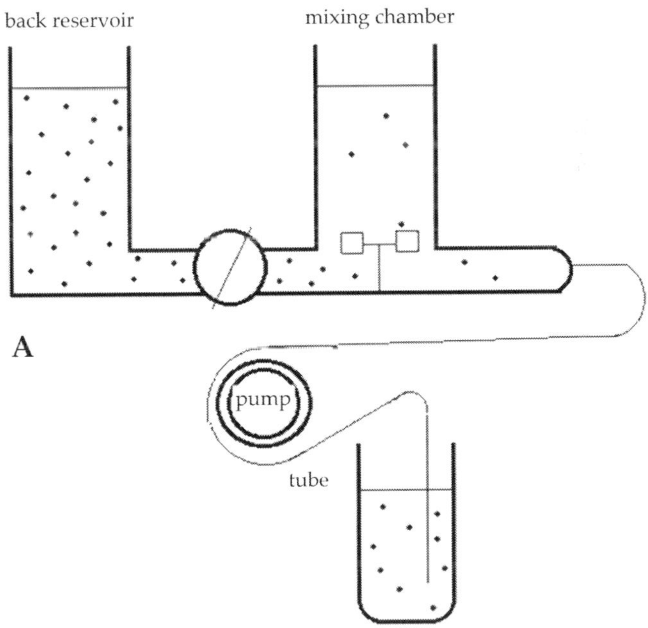

Figure 37: Formation of a continuous density gradient with a density gradient maker using the low density method.

The lower density solution is initially placed in the mixing chamber and added first to the bottom of the centrifuge tube. As the process continues, the denser solution which is in the back reservoir is slowly added to the mixing chamber thereby increasing its density. Thus in this process the initial low density solution is continuously replaced from the tube bottom by a more dense solution thereby creating a gradient.

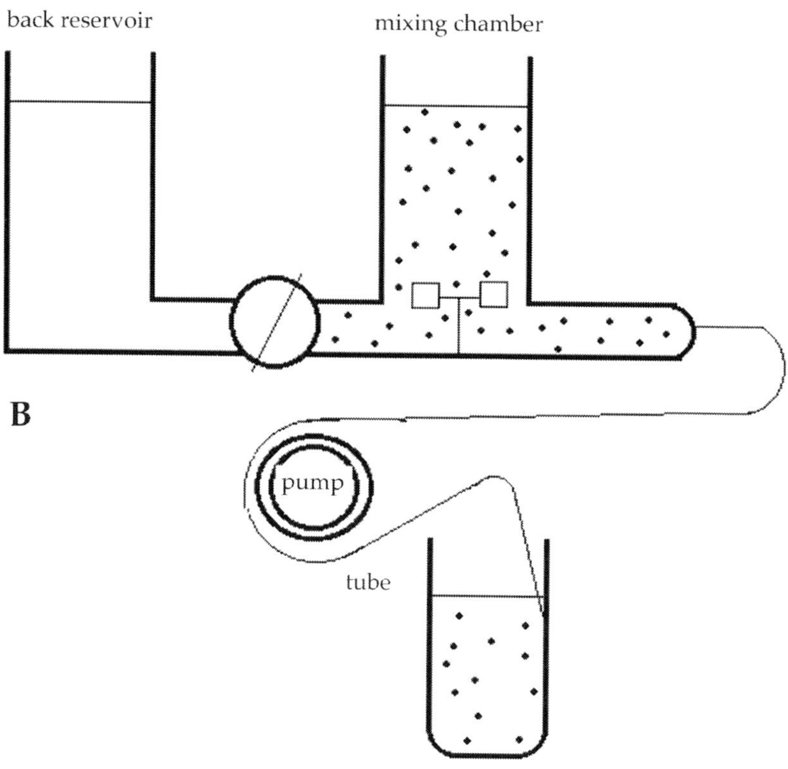

Figure 38: Formation of a continuous density gradient with a density gradient maker using a higher density method.

In this method the solutions are reversed with the denser solution being placed in the mixing chamber. The solution is also added from the tohe centrifuge tube whilstwhile in the previous method it was added from the bottom. Less dense solutions are subsequently added from the top thereby creating a gradient.

CHOOSING A SUITABLE DENSITY GRADIENT MEDIUM

The main function of density gradient centrifugation is to separate particles. This can happen either on the basis of their buoyant density or their rate of sedimentation. For rate-zonal separations, the function of the gradient is to provide a medium of viscosity which improves particle resolution while at the same time stabilizing the column. For isopycnic separations, the important feature is that the maximum density of the gradient medium is higher than that of the densest particles. A wide range of compounds have been used to form gradients for rate-zonal and isopycnic separations such as sucrose, glycerol and caesium chloride. Depending on the range of compounds used, the choice of density gradient will vary depending on the type of separation (Rickwood, 2001). An ideal density gradient medium should have the following properties:

- It should be sufficiently soluble such that the required range of densities can be produced during centrifugation.
- It does not form solutions with high viscosity in the desired density range.
- It does not affect osmotic abilities when the particles to be separated are osmotically sensitive.
- It should be adjustable so that a compatible pH and ionic strength with that of the particles being separated can be achieved.
- The density gradient should not affect the biological activity of the sample.
- It should be nontoxic and inflammable.
- It should not be metabolized by cells.
- It should not interfere with assay procedures or react with the centrifuge tubes.
- It can be used as a measure of concentration.
- It is easily removed from the purified product.
- It is autoclavable.
- It is cost effective and readily available.
- It is stable in solution.

- It should be inert and not react with other biological material, the centrifuge or component parts.
- It should not absorb light at wavelengths generally used for spectrophotometric readings.

While no ideal gradient material exists and no single compound can satisfy all of the above criteria, the choice of gradient material chosen should depend on the nature of the particles to be separated. Most media should be capable of producing the range of densities required and be easily removable from the desired particles of interest. A variety of different compounds have been developed for use as density gradient media (Table 10) in order to enhance the separation process and to overcome difficulties associated with osmolality and viscosity.

Table 10: Density gradient media types and respective applications

GRADIENT MEDIUM TYPES	PRINCIPLE USES
POLYHYDRIC ALCOHOLS	
Sucrose	Organelles, membrane vesicles, viruses, proteins, ribosomes, polysomes
Glycerol	Mammalian cells, proteins
Sorbitol	Non-mammalian subcellular particles
POLYSACCHARIDES	
Ficoll	Mammalian cells, mammalian subcellular particles
Polysucrose	Mammalian cells, mammalian subcellular particles
Dextrans	Mammalian cells, mammalian subcellular particles
INORGANIC SALTS	
CsCl	DNA, viruses, proteins
Cs_2SO_4	DNA, RNA
KBr	Plasma lipoproteins
IODINATED GRADIENT MEDIA	
Nycodenz, Histodenz	Mammalian cells, organelles, membrane vesicles
Iodixanol	Mammalian cells, organelles, membrane vesicles, viruses, plasma lipoproteins, proteins, DNA

COLLOIDAL SILICA MEDIA	
Percoll	Mammalian cells, organelles, membrane vesicles

(adapted from Sigma-Aldrich, 2011)

COMMONLY USED DENSITY GRADIENT MEDIA

Overall, gradients can be either shallow or steep. Shallow gradients consist of the two initial solutions having similar densities while steep gradients consist of the two initial solutions having very different densities. Examples of commonly used gradient materials include the salts of alkali metals such as caesium and rubidium chloride, small neutral hydrophilic organic molecules such as sucrose, hydrophilic macromolecules such as proteins and polysaccharides, a number of miscellaneous compounds such as colloidal silica and non-ionic iodinated aromatic compounds such as metrizamide. Each compound has their own benefits and should be selected depending on the type of particles that are required for separation by centrifugation (Table 8). For example, sucrose solutions are most commonly used for rate-zonal centrifugation despite being very viscous at densities greater than 1.1 to 1.2 g cm^{-3} while glycerol is most commonly used for the separation of enzymes and Ficoll is used for the separation of whole cell and subcellular organelles.

In addition, the effect of osmolality is of high important and should be taken into consideration when choosing a suitable density gradient medium. High osmolality solutions will remove water from the interior of membrane bound particles as well as remove water that is bound to macromolecules such as DNA. This loss of water will reduce the size of cells thereby increasing their density and ultimately affecting their buoyancy and rate of sedimentation during centrifugation. This osmotic effect on cells and macromolecules may be reversible ;showever, this kind of error should be taken into consideration before centrifugation so as to be avoided. There are five main classes of density gradient medium, namely, polyhydric (sugar) alcohols, polysaccharides, inorganic salts, iodinated compounds and colloidal silica.

Polyhydric Alcohols

Polyhydric alcohols are also known as sugar alcohols, or polyalcohols, and are a class of organic compounds generally derived from sugars such as sucrose, glycerol and sorbitol. These organic molecule are small, hydrophilic and are considered non-ionic gradient media. Sucrose gradients are widely used for rate-zonal separation of macromolecules and for the isopycnic separation of viruses and cell organelles since it meets most of the criteria for an ideal medium. It is biologically inert, stable, and relatively cheap. In addition, it is well characterised in terms of its concentration, viscosity and density which in turn makes it easier to develop and adapt methods for the separation of various particles. However, while sucrose may have little effect on intermolecular bonding, the high osmotic pressure may result in shrinking of enveloped viruses. In addition, concentrated solutions tend to be viscous and hypertonic and using reagent grade sucrose may lead to contamination with RNAses or heavy metals thus rendering it unsuitable for DNA and RNA centrifugation and purification. Other sugar solutions such as glycerol and sorbitol have been used effectively in rate zonal centrifugation and are less dense and can be used to preserve the activity of certain enzymes.

Table 11: Commonly used gradient material and maximum densities at 20 °C in water as a solvent

Materials	**Maximum Density**
High viscosity, low density	
Ficoll	1.17
Glycerol	1.26
Silica sols	1.30
Sucrose (66%)	1.35
Albumin	1.35
Sucrose	1.37
Diodon	1.37
Sorbitol	1.39
Meltrizamide	1.46
Low viscosity, low density	
Lithium chloride	1.33

Sodium formate	1.40
Potassium acetate	1.41
Rubidium chloride	1.49
Potassium formate	1.57
Low viscosity, high density	
Lithium bromide	1.83
Rubidium formate	1.85
Cesium chloride	1.98
Cesium acetate	2.00
Cesium formate	2.10

(Adapted from Griffith, 2010)

Polysaccharides

High molecular-weight polysaccharides such as Ficoll, dextran, glycogen and polysucrose can also be used as a gradient medium. While sucrose solutions may have the problem of high osmotic strength, polysaccharide solutions do not penetrate intact biological membranes and have a lower osmolality making them useful during separations of sensitive, membrane bound viruses. While polysaccharides present some advantages for centrifugation, there are some disadvantages too. Due to the size of the polysaccharides, they cannot be removed from the sample by simple ultrafiltration; in this case dilution and high-speed centrifugation are required. The most common polysaccharide medium used for density gradient centrifugation is Ficoll which is produced by the polymerization of sucrose molecules with epichlorohydrin. Ficoll solutions below 20 % (w/v) have a density of 1.07g/cm^3 and are considered osmotically inert. However, since many polysaccharide solutions have slow diffusion properties, it may be beneficial to form linear gradient solutions using a gradient mixer thereby ensuring that more stability is maintained (Lawrence and Steward, 2010).

Metal Salts

Alkali metal salts generally consist of concentrated heavy metal salts such as caesium chloride, caesium sulfate, sodium iodide, sodium bromide and the lithium and rubidium salts are commonly used for making isopycnic gradients. Metal salts can provide some of the densest gradients available. They have a low viscosity, and their concentration in solution is easily measured by refractive indexes (Lawrence and Steward, 2010). The major disadvantage with using these salts however, is based on their effect on biological activity. These salt solutions have high ionic strengths which ultimately disrupt protein-protein and nucleic acid-protein bonds. They also have high osmolalities which in turn can affect particle hydratiforeTherefore, these metal salts should be limited to use where high yield and purity are needed but where biological activity is not required (Steward et al. 2000). Caesium chloride and caesium sulfate are the most widely used heavy metal salts with gradient densities of up to 1.91 g/cm^3 (Majekodunmi, 2015).

Iodinated Density Gradient Media

Ionic gradient media have higher densities and lower viscosities than other density gradient we theirhowever, their use is restricted due to the high osmolalities and ionic strength which affect the hydration of osmotically sensitive particles. In addition when compared to polyhydric alcohols, iodinated gradient media have much lower osmolalities and viscosities than sucrose regardless of the concentration. Examples of commonly used iodinated organic compounds include Nycodenz, OptiPrep, and metrizamide which form excellent density gradient media that provide biological inertness, a wide density range, and low viscosity. These compounds are heat stable, autoclavable and of minimal ionic strength. The iso-osmotic nature helps to minimize any morphological artefacts which may be useful if the structure of a virus is to be investigated by electron microscopy. These advantages make them an excellent choice for isopycnic separations of viruses.

Colloidal Silica

Common examples of colloidal silica suspensions include Percoll and Ludox which are non-ionic media that can be used to rapidly generate self-forming gradients. While colloidal media cannot be effectively filter sterilized, they may be autoclaved before being adjusted for osmolality and can be used over a wide pH range. Colloidal silica media are not true solutions but are colloidal suspensions of silica particles coated with polyvinylpyrolidone (PVP) and are most often used in the isopycnic separation of cells, viruses and organelles. The PVP minimizes the particle interactions with biological material and stabilizes the colloid while the osmotic strength is extremely low and changes little with density. The osmolality can be adjusted by adding appropriate amounts of sucrose or buffer solution. The most widely known colloidal silica medium is Percoll which is commonly used for cell separations because the suspension of colloidal silica can be prepared in almost any buffer required to maintain cell viability. The main limitation is that the sample particle size must be larger than that of the colloidal silica particles, otherwise the particles of silica will pellet before the sample bands. Viruses can also be separated in gradients of colloidal silica but removal of the gradient medium from the viruses following centrifugation can be difficult since both the viruses and the medium are colloidal (Pertoft et al. 1967). In addition, silica particles may begin to pellet before smaller viruses have time to form distinct bands.

Gradient Shapes

The shape of a gradient is defined as the change in the density of a solution in relation to the physical length of the centrifuge tube or the distance from the centre of rotation (Figure 39). This usually forms a function of the gradient material and its concentration as well as the rate of centrifugation. There are five principal types of gradient shapes or systems that are used in rate-zonal separations (Table 12).

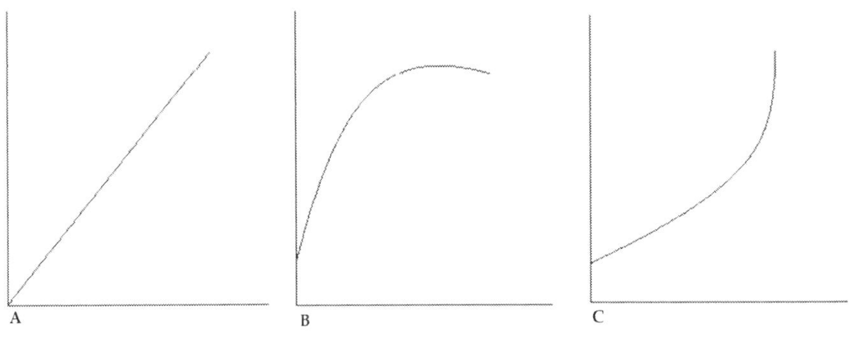

Figure 39: Examples of the shapes of different gradient profiles. Continuous gradients may be A) linear, B) convex or C) concave and can be used for rate zonal or isopycnic separations.

These shapes can be modified and adapted to the specific conditions required and is usually determined by the concentration profile along the tube. The choice of gradient shape depends on the kind of separation required and the relative quantity of the sample to be separated. The slope of the gradient will determine the resolution with small quantities of particles (about < 0.1 mg per ml of gradient) being safely managed on those gradient shapes designed for high resolution. Linear gradients are the simplest shape for isopycnic gradient separations while hyperbolic gradients should be used for larger quantities of samples.

Table 12: Types of gradients used in rate zonal centrifugation

TYPE	GRADIENT CHARACTERISTICS *
Linear	Density solution is linear to distance
Isokinetic	Density solution is convex (sometimes linear) to distance. All particles of the same density sediment at a constant rate at all distances from the centre of rotation
Step /discontinuous	Density solution forms discrete zones in relation to distance. There are relatively large differences in density and concentration.
Shallow	Density solution changes gradually in relation to distance.
Sharp / steep	Density solution change rapidly in relation to distance.

* These characteristics are listed as a function of distance along the centrifuge tube.

Ref: Adapted from Spinpro Software from Beckman Instruments

The choice of gradient material and the shape of the gradient can in turn control the properties of the sedimenting zone of particles. Previous initial studies conducted by Harrison and Nixon (1959), de Duve et al (1959) and subsequently by Martin and Ames (1961) showed that a gradient shape can be selected with a specific purpose in mind. For example, an isokinetic gradient can be selected to control the distance that a protein moves in a gradient and for making this proportional to the sedimentation coefficient of the protein. This in turn is equivalent to making the sedimentation rates of any one protein constant throughout the gradient. A step gradient on the other hand consists of segments of homogenous solutions of increasing density (Figure 40). A pure step gradient exists for only a few moments after being formed before diffusion of the solute occurs and replaces the sharp step with a steep but smooth slope (dotted line in Figure 40).

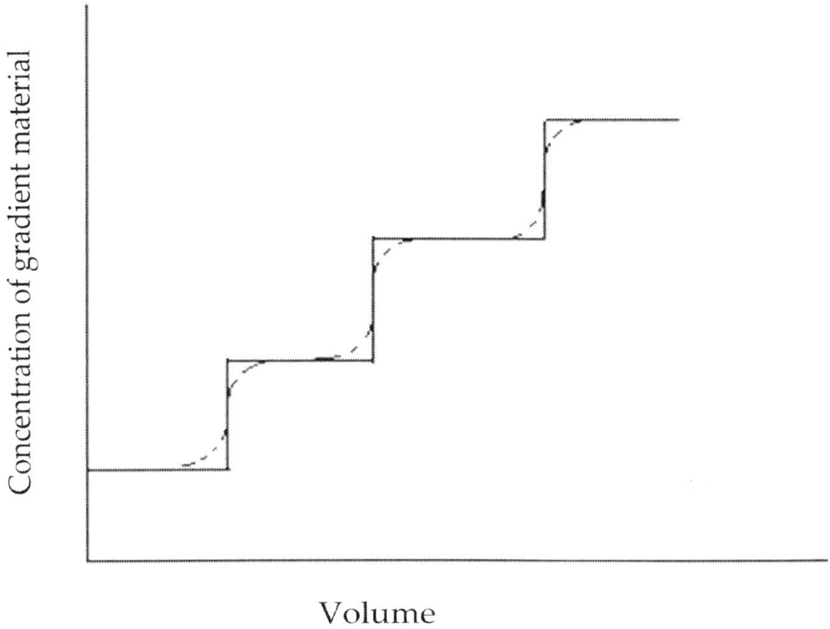

Figure 40: Simple step gradient constructed by loading discontinuous volumes of solution with apparent increases in solute concentration (solid lines). The sharp steps are momentary and begin to diffuse quickly into rounded steps (dotted lines).

SAMPLE COLLECTION AFTER GRADIENT FRACTIONATION

Once a centrifugation run has been completed, tubes should be handled gently and with as little mechanical movements as possible. Samples should also be recovered as quickly as possible to prevent re-diffusion of separated fractions. There are a number of methods that can be carried out for recovering samples from gradients ; however, the specific method chosen will depend on the nature of the separation, the type of centrifuge tube being used and the equipment currently available. In addition, while most recovery methods can be successfully completed with the use of little or no equipment, the resolution of collected particles in a band and subsequent reproducibility of the method in question can be improved with the use of specialized gradient-harvesting equipment. There are a range of methods that can be used for the collecting of an individual band or fractions of particular particles.

The Side Puncturing Method

If the desired band of particles can be easily seen in the gradient solution then collection or harvesting by direct puncturing of the tube can be done. Depending on the material of the centrifugation tube, it may be punctured with a needle and syringe just below the band of interest. The hole of the needle should be pointed upwards and clear adhesive tape can be placed around the desired area to be punctured in order to maintain a seal around the needle (Figure 41). The material is then harvested by drawing the liquid slowly into the syringe until the band is no longer visible in the centrifuge tube. If a sealed tube is used, another puncture must be made near the top of the tube to relieve any pressure. One of the major difficulties associated with this method is efficiently making a puncture through the side of the tube without any sudden disturbances. The tube itself may deform from the change in pressure when forcing the needle through and air bubbles may be introduced through the punctured hole which can in turn rise through the gradient causing some mixing of the band of interest and surrounding solution. In addition, if the needle is pushed directly through the wall of the tube, a small piece of plastic may be removed from the wall of the tube and block the opening of the needle. For this reason the needle should be twisted as the tip penetrates the wall. In addition, the use of a needle could also endanger the user as it may slide quickly and penetrate the opposite side of the tube, and potentially the users finger.

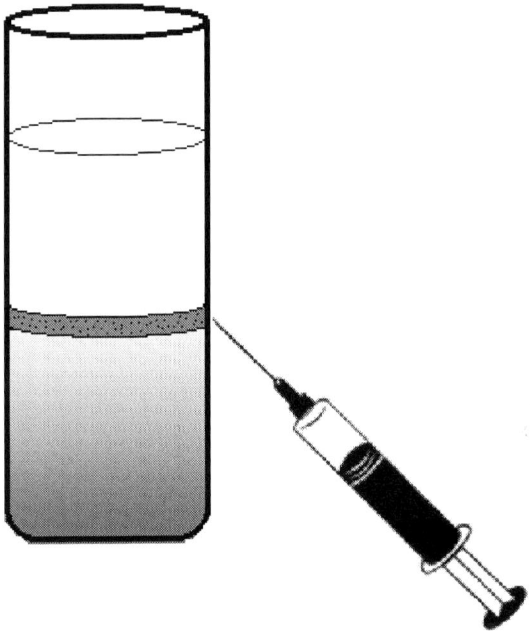

Figure 41: Targeted harvesting of a single band of interest from a density gradient by puncturing the side of the centrifuge tube with a needle.

Gradient Fractionation

Fractionation is one such method that is used to recover bands and can be done via a range of automated or hand-driven methods. Relevant precautions should be taken to avoid loss of resolution of the particles (Yamazaki, 2000). These precautions include ensuring that 1) all operations should be carried out such that disturbance of the gradient is minimzed, 2) the volume of tubing running from the gradient to the fraction collector should be minimzed 3) contamination of the fractions to be recovered is avoided and 4) fractionation of the gradient should be done at a slow flow rate especially if viscous materials are used within the gradient.

Gradient Fractioning using a Direct Unloading Method

The simplest method of collecting fractions from a gradient is to aspirate it from the top of the gradient using a pipette. By keeping the pipette tip as close as possible to the meniscus during each aspiration, different fractions can be collected and removed starting with the least dense fraction being removed first (Figure 42). This method can also be done using an automated device such as a machine referred to as an Auto Densi-Flow. In this automated apparatus, a tip with side holes is attached to a motorized mount that moves up and down. A sensor maintains the inlet holes of the tip just below the meniscus and a peristaltic pump drives the aspiration. Another method is to insert a tube at the bottom of the centrifuge tube or gradient and draw the liquid out, starting with the densest first. This is however not recommended since the gradient is inverted and as it moves up the tube may lead to mixing and loss of band resolution.

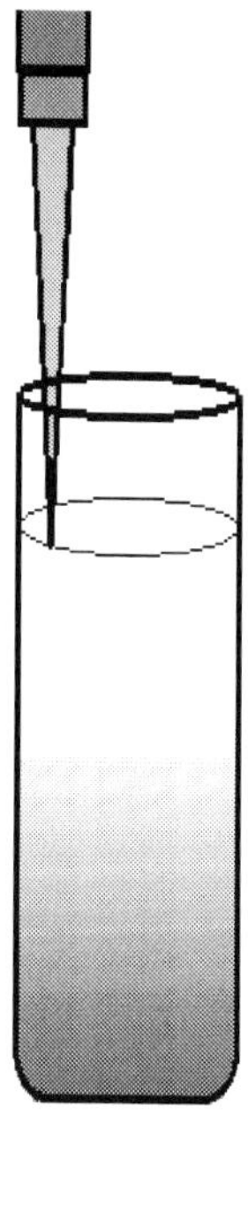

Figure 42: Collection of gradient fractions from the top of a centrifuge tube using a pipette. The tip of the pipette should be kept as close to the meniscus of the liquid as possible.

Gradient Fractioning using the Bottom Puncture Method

The bottom puncture method is one of the most effective methods for harvesting fractions from a gradient and involves puncturing of a hole in the bottom of the centrifuge tube followed by harvesting the desired fraction in drops from the bottom. This is known as the gravity drip method and is one of the most preferred methods because it is simple and requires no special equipment. The centrifuge tube is firstly secured above a benchtop using clamps that apply just enough force to keep it from slipping. This is followed by making a hole at the top to allow the pressure to reduce and equalize before harvesting can begin. Pressure differences may result from squeezing of the tube into the clamp or from a difference between the current temperature of the liquid in the tube and its temperature when the tube was sealed. Care should be taken if the bottom hole is punctured first as there may be an uncontrolled flow until the pressure is equalized and mixing of the separated fractions may occur. Fractions are then collected by dripping into a series of tubes.

An alternative method for open top tubes is to use a flow rate method which involves piercing the bottom or near the bottom of the tube with a needle attached to a stopcock (Figure 43). A twisting motion is generally used to ease piercing of the needle through the wall and to avoid plugging the needle with plastic. The position of the needle should be adjusted such that the opening is as near to the bottom of the tube as possible. The stopcock can then be opened to give the desired flow rate. In addition, instead of relying on gravity, the sample may also be drawn from the bottom of the tube via a peristaltic pump to further control the flow rate or the gradient may be pushed out of the bottom by displacement from above. In the latter case, the needle used to pierce the top of the tube remains in place and is connected to tubing passing through a peristaltic pump. Either air or pure water is then slowly pumped into the top of the tube to control the flow rate through the bottom needle and stopcock assembly.

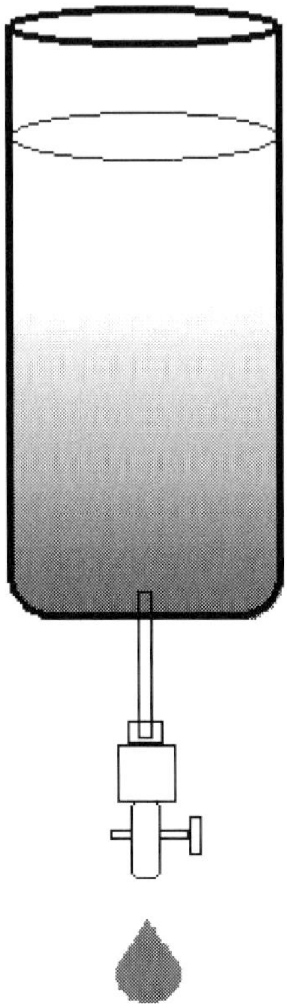

Figure 43: Gradient fractionation using a gravity drip collection method with control of pressure and flow through a bottom stopcock.

Gradient Fractionation using an Automated Method

Another method for removing gradients from open top tubes is to use a device known as a Piston Gradient Fractionator. This device has a piston that seals against the walls of the tube. When the piston is pressed down, the liquid escapes through a hole in the tapered piston head and travels through the tubing to the collection point (Figure 44). This device allows for the harvesting of fractions as thin as 300 μm and aids in preventing cross-contamination between fractions by rinsing and drying of the tubing between sampling. An additional advantage provided by this device is the improved visual detection of bands within a gradient solution. The sample holder maintains and supports the centrifuge tube within a bath of water which allows the centrifuge tube to be transparent, thereby providing illumination from below.

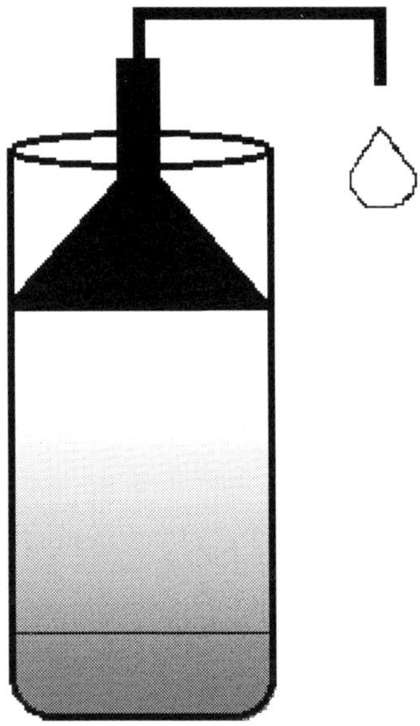

Figure 44: Gradient fractionation through top unloading of fractions through a plunging, tapered piston.

Fractionation Collection using the Displacement Method from below

In this method, a liquid that is denser than the densest part of the gradient is pumped into the bottom of the tube. This newly pumped in liquid displaces the liquid out the top of the tube through a specialized cap. The dense solution can be added through a needle which pierces the bottom of the tube or by inserting a narrow tube to the bottom of the gradient from above (Lawrence and Steward, 2010).

MEASURING THE DENSITIES OF SEPARATED FRACTIONS

Being able to determine the density of a sample is important as it aids in verification of the density of stock solutions, estimation of the buoyant densities of particular particles of interest and even the determination of the gradient's final shape. Refractometry is one of the quickest methods available as it allows for accurate measurements, ease of operation and precision. This method however does require specialised equipment and cross-contamination may be a possibility since all samples are known to come into contact with the same surface of the utilised equipment. In addition, if the gradient medium was initially prepared in a buffer then corrections of the refractive index must also be done especially if a mixed gradient is used.

A more direct method for measuring fraction densities is to determine the mass of a known volume of sample using a balance and micropipette. The technique used is simple and easy. The following basic steps are employed:

- A tube containing a fraction is placed onto a balance.
- The balance is tarred.
- A fixed volume is removed with a pipette.
- That removed volume is retained in the pipette while the mass is recorded.
- The sample is then dispensed back into the same tube.

The advantages associated with this method are such that no volume is lost, nor is it contaminated with other samples. Repeat measurements can be done quickly and efficiently using the same fraction. In addition, the type of pipette chosen is important. Regular air-displacement pipettes tend to under collect volumes with increasing density which may result in the underestimation of fractions. It is recommended that positive displacement pipettes be used or if air-displacement pipettes are used then the error should be corrected for. The samples should be well mixed before taking a sample measurement and the temperature of the solution should also be taken into consideration. When centrifuging solutions it is important to pay attention to the centrifugation speed, temperature and concentration of the solution. Instruction manuals for rotors are provided and these generally contain the maximum safe densities for a given speed as well as run temperatures; however, these are generally measured at 20 °C. This means that if operation near the maximum allowable density is going to occur, then measurements of the initial density should be made close to 20 °C or a correction factor should be applied. For example, if a gradient was run at 4 °C but a virus is generally found in a fraction whose density was measured at 24 °C then one will have to account for the thermal expansion from 4 °C to 24 °C in order to determine the density at which the virus will have banded (Lawrence and Steward, 2010).

APPLICATIONS OF DENSITY GRADIENTS

The main purpose of density gradients is to separate and purify a range of cells, organelles or macromolecules like nucleic acids and proteins (Schumaker, 1967; De Duve, 1971; Birnie, 1972). They are also used to measure the buoyant densities and sedimentation coefficients of particles, to estimate size of proteins and nucleic acids and to investigate the effects of chemical, physical or biochemical treatment of sample materials. Density gradient centrifugation can be utilised in a range of areas such as for the separation of living cells and cellular organelles, applications in marine viral ecology and stem cell research and in the separation of microalgae. For further information refer to the applications section further on.

CHAPTER 6

CENTRIFUGES

TYPES OF CENTRIFUGES

The overall design and size of a centrifuge will usually depend on its particular applications. The main operating principle is that all centrifuges will contain a central motor that spins a rotor which contains the samples with particles that are required to be separated. Particles that are required to be separated are usually suspended in a solution or liquid buffer which is then contained in specific centrifuge tubes. Many different types of centrifuges are commercially available (Table 13). These include large capacity low-speed preparative centrifuges; refrigerated high-speed preparative centrifuges; analytical ultracentrifuges; large-scale clinical centrifuges and small-scale laboratory microfuges. There are generally 4 main types of centrifuges: 1) clinical benchtop centrifuges; 2) high speed refrigerated centrifuges; 3) continuous flow centrifuges and 4) ultracentrifuges. Ultracentrifuges can further be classified into preparative and analytical ultracentrifuges. The most obvious differences between centrifuges are the maximum speed at which biological specimens can be subjected to, the presence or absence of a vacuum, the presence of refrigeration or general manipulation of the temperature during a centrifugation run and the capacity of individual centrifugation tubes that fit in the rotor.

Low Speed Centrifuges

Low speed centrifuges have maximum rotor speeds of less than 10 000 rpm. This means that rotors are not required to be operated in a vacuum and excess heat generation is minimal. In addition, many of these low speed centrifuges have a temperature control facility and a sensor that will detect any imbalances when the motor is running. Should any kind of imbalance occur, this detection system will cut off the power to the drive mechanism and stop spinning of the rotor. Low speed centrifuges are commonly used for the initial processing of biological samples such as cells and organelles that contain target proteins as well as to separate serums or plasma from red blood cells. They are commonly used to harvest and purify sub cellular components like nuclei, large mitochondria and plasma membrane fragments.

High Speed Centrifuges

High speed centrifuges are usually used for more sophisticated biochemical applications and can achieve higher speeds of up to 30 000 rpm and RCF's of up to 120 000 × *g*. Since these instruments operate at such high speeds, refrigeration systems are required to overcome the heat generated by the friction produced during spinning of the rotor. Rotor chambers in most instruments can reach temperatures of up to 40 °C hence the need for refrigeration. In addition vacuum systems must also be incorporated. These centrifuges are used in the separation of a number of cellular components as well as in the isolation and purification of viruses, soluble proteins and microorganisms.

Small Benchtop Centrifuges

These are the simplest type of centrifuges available and are used in many laboratories for routine work such as sedimentation of cells as well as plasma and serum separation. The maximum speed of this type of centrifuge ranges between 4 000 to 6 000 rpm. In addition, because they are used at such low speeds, they can be used with or without refrigeration and can hold up to 100 tubes depending on the diameter of the rotor.

Clinical Benchtop Centrifuges

These are compact benchtop models which are specifically designed for use in hospitals and clinics and which spin blood collection tubes and urine samples at very low speeds for diagnostic examinations. Most of these centrifuges run at speeds 3000 × g RCF or below. Generally these models can hold up to 28 tubes and a sample volume of between 3 to 15 ml in centrifuge tubes.

Micro Centrifuges

Micro centrifuges are commonly found in every laboratory and are similar to benchtop models. Despite being compact and small, these units provide RCFs of up to approximately 21 000 × g and are usually used in the isolation of plasmids, DNA and RNA. These models utilise ordinary PCR tubes or microcentifuge tubes that can hold samples of up to of 2 ml volumes.

Continuous Flow Centrifuges

This is one kind of high speed centrifuge where the rotor is specially designed to allow for the sedimentation of particles through a continuous flow of medium. Usually a large volume of medium is passed through. Here the cells or particles are sedimented against the wall and the excess medium or liquid is released through the exit tube.

Ultracentrifuges

This is most sophisticated of all the centrifuges because of the high speeds attained. These high speeds also denote the need for an efficient chamber refrigeration and high vacuum system to reduce the friction generated. The ultracentrifuge consists of a refrigerated low-pressure chamber containing a rotor which is driven by an electric motor that is capable of reaching very high speeds. Rotational speeds may reach up to 100 000 rpm for floor models and 150 000 rpm for bench top models resulting in centrifugal speed forces of up to 800 000 *g* to 1 000 000 *g*. There are two kinds of ultracentrifuges, namely the preparative and analytical ultracentrifuge.

Preparative Ultracentrifuges

This instrument allows for the isolation and purification of macromolecules or cell organelles. It operates in refrigerated conditions under a vacuum to avoid frictional resistance of the rotor caused by the spinning air. The whole system employs a continuous monitoring system which monitors the rotor temperature through the use of a temperature sensor. Preparative ultracentrifuges can attain a maximum speed of up to 80 000 rpm and can produce a centrifugal force of 600 000 *g*.

Analytical Ultracentrifuges

The analytical ultracentrifuge is used for determining the molecular weight as well as the hydrodynamic and thermodynamic properties of macromolecules. No other technique is capable of providing the same range of information with such precision and accuracy. An analytical ultracentrifuge has one or several optical detection systems which allow for the observation of the fractionation process while the sample is being centrifuged. This instrument has many applications such as allowing for the molecular weight, purity and shape of macromolecules to be seen. It can run at speeds of up to 80 000 rpm with a centrifugal force of about 500 000 *g*. It contains a special optical system to determine the concentration and distribution within the sample during centrifugation. Monitoring in this system is done either through the ultraviolet absorption system or by noting the differences in the refractive index. If the concentration is uniform, light passes through it without any deviation but if the light passes through a solution of different density zones, it is refracted at the boundary between these zones. By measuring the refractive index between the reference solvent and the solution, the concentration of solute at any point can be measured. Furthermore the rate of sedimentation can be measured by taking photographs of the moving boundaries of sedimenting particles. To separate particles of different densities, they are labelled with H^3 or C^{14} and then centrifuged in a sucrose density gradient. Particles labelled with H^3 have a greater mass and sediment faster than those labelled with C^{14}.

Table 14: Classes of centrifuges and their separation applications

Parameters	Low speed	High speed	Ultracentrifuge
Speed ranges (rpm x 10^3)	2-6	18-22	35-120
Maximum RCF (x10^3)	8	60	700
Bacteria	-	Yes	Yes*
Animal and plant cells	Yes	Yes	Yes*
Nuclei	Yes	Yes	Yes*
Precipitates	Some	Most	Yes*
Membrane Organelles	Some	Yes	Yes
Membrane fractions	Some	Some	Yes
Ribosomes / polysomes	-	-	Yes
Macromolecules	-	-	Yes
Viruses	-	Most	Yes

*Separation can be done but this type of centrifuge is not usually used for this purpose (adapted from Sigma-Aldrich, 2011)

Industrial Centrifuges

Centrifuges are known to service many manufacturing industries on a large scale. They serve a range of applications related to the separation of components based on their densities and involve the separation of insoluble particles from supernatant liquids as well as the extraction of dissolved substances from one immiscible liquid to another. All steps such as the blending of liquids and transfer of solute particles and the separation of immiscible phases are carried out in the same machine. Over the last few years there has been a growing need for industrial centrifuges resulting in a variety of designs being developed. These designs however can generally be divided into two main classes namely sedimenting centrifuges and filtering centrifuges (Table 15).

Industrial centrifuges operate on two basic principles. The first is that when centrifugal acceleration is exerted on suspended particles, it will cause them to move outwards where solids particles will then collect on the wall of the rotating body. A second principle of operation is based on density differences between two immiscible liquids such as fat separation in the dairy industry. Sedimenting centrifuges operate in such a way that solids are transported to the outer most wall of the centrifuge bowl where it is either collected on a separate screen or the wall itself.

Liquids are removed via the close packing of particles from the initial mother suspension. With filtering centrifuges, the solids are transported to the surface and usually trapped on a filter where the liquids drain through (Beveridge, 2000). An additional parameter that is important in classifying industrial centrifuges is whether the machines are fed continuously or in batch modes. Sedimenting centrifuges are further divided into batch and continuous models while filtering centrifuges are divided into batch and continuous fed designs.

Table 15: Classification of industrial centrifuges according their principle of operation

	CONTINUOUS FEED	**BATCH FEED**
Sedimentation centrifuges	Solid bowl decanter	Imperforate basket
	Tricanter	Solid bowl
	Disk centrifuges (intermittent discharge, nozzle discharge, hydrocarbons) Hydrocyclones	Tubular
Filter centrifuges	Pusher centrifuge (single or double stage)	Vertical basket
	Screen / scroll	Peeler
	Screen bowl decanter	

(adapted from Majekodunmi, 2015)

Sedimentation Centrifuges

The operating principle of sedimentation centrifuges is to collect particles against the centrifuge bowl wall. This sedimentation is based on the centrifugal speed, particle size and density, centrifuge radius and fluid viscosity with the rate of sedimentation increasing as either of these factors increase. The size, shape and design of the centrifuge itself will also determine the overall performance of the run which usually depends on the purpose of centrifugation, purity of separated phases and extensiveness of separation between the liquid and solid phase. Performance can also be measured by several other criteria such as moisture content, total solids recovery, flocculent concentration, yield, particle purity, power consumption and overall maintenance of equipment.

Separation Properties with Sedimentation Centrifuges

Separation performance with sedimentation centrifuges is often enhanced by increasing the diameter of the particles to be sedimented. This is usually done through the use of certain coagulants or flocculants (Table 16) and by changing certain parameters. Coagulants and flocculants will destabilize particles and cause a form of attraction between them, thereby causing them to agglomerate together and form larger particles. In addition, changes in surface charge, hydrophobic or hydrophilic nature as well as water-binding abilities will also affect the sedimentation properties of particles. A wide range of coagulants or flocculants are readily available for use with various particles to be sedimented. These should be chosen after careful analysis of the chemical nature of the particles to be separated (Beveridge, 2000).

Table 16: Commonly used coagulants and flocculants for sedimentation centrifugation

Metal salts	Aluminium salts $Al_2(SO_4)\cdot 16H_2O$
	Ferric iron salts $Fe_2(SO_4)_3\cdot 9H_2O$

Natural flocculants	Starch
	Gums
	Tannin
	Alginic acid
	Sugar or sugar acid polymers
	Polyglucosamine
Synthetic flocculants	Polyacrylamides
	Polyamines
	Cellulose deriavtives

(Adapted from Whittington, 1990)

Filter Centrifuges

These centrifuges operate by sedimenting particles on to a screen which usually consists of a porous membrane or filter cloth to collect and hold the solids. The presence of holes or slots in the membrane or cloth also allows for the liquid portion to flow through and subsequently move away.

Separation Properties with Filter Centrifuges

For effective separation of large volumes containing high concentrations of solids, filter centrifuges require that solids should be approximately 100 – 200 μm in diameter and easily allow liquids to drain past. Feeds containing high solid concentrations should be used together with pre-treatment concentrators such as settling tanks.

Continuous Feed Centrifuges

Continuous feed centrifuges form second and third generation centrifuges and consist of a continuous discharge of both liquid and solids from the centrifuge bowl or basket. Common examples include conical and pusher centrifuges. The main purpose of continuous feed centrifuges are to allow for the continuous feeding of solutions and the continuous separation of the solids and liquid phases. These are usually used when separations of extremely high volume materials are required thereby allowing runs to proceed without the need for stopping to decant and refill the machine before starting again. In this way a great deal of time is saved due to the reduction in sample handling times and not having to wait for rotors to accelerate and decelerate between runs.

Decanter Centrifuges

Decanterscentrifuges are designed in such a way that they are the only sedimenting models that can handle a significantly large amount of solids right from the start. The main purpose of decanter centrifuges are to separate solid materials from liquids in slurry-like solutions andetherefore, has become an important component of the wastewater industry as well as the chemical, oil and food processing industries. These centrifuges consist of alcylindrical bowl and helical extraction screw which fits into the inner outlines of the bowl creating a form of clearance (Figure 49). The main operating principle is based on that of gravitational separation whereby through the force of continuous rotation, heavier particles will congeal at the bottom causing liquids to be forced upward creating a clear separation between the two phases. A decanter centrifuge usually contains a bowl that rotates at high speeds. The slurry is pumped in through a connecting pipe and onto the internal conveyor which is moving in the same direction as the bowl but at a different velocity. From here, the slurry is pushed into the bowl area which is rotating at high speeds resulting in an induced gravitational force. This high speed rotation results in the separation of solid material from that of liquids. The solid material is pushed upwards where it is subsequently discharged via a nozzle while the liquid is collected via a different output area (Muze Development, 2016). One of the main advantages of decanters is the continuous separation of a range solid concentrations in varying feedstocks. While common two phase machines that offer separations of the solid and liquid phases of mixtures are available there are also those models that offer the separation of three phases such as an aqueous phase, an oil phase and a suspended solids phase. These are known as tricanter centrifuges. Decanters are most often used in the wastewater industry due to their continuous and effective separation of solid and liquid phases while being able to handle high solid content feeds in a continuous cycle. For similar reasons, they have also been used for dewatering coal and in the separation of mineral rich slurries within the mining industries.

Tricanters are often used for refining vegetable oils, producing fat and milk as well as in the recovery of waste oil. Since tricanters are able to efficiently separate out the mixtures into three phases, the recovered oil that is separated can then be refined and sold to other industries that may need it. In addition, it may also be used in the production of fruit and vegetable juices whereby the pulp oil, juice and seed containing solids are separated. Despite offering such effective separations, decanter centrifuges do not always offer complete clarification of liquids as some suspended solids may remain.

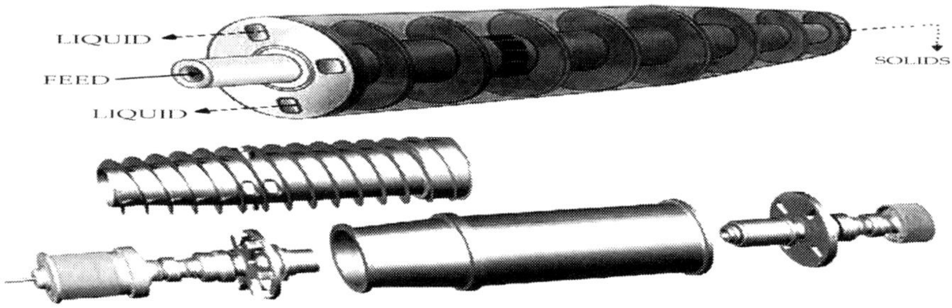

Figure 49: Schematic diagram of a decanter centrifuge showing the flow of initial feed and separated liquid and solid sections. Reproduced with permission from Relex Process Consultancy Ltd.

Benefits of the Decanter

One of the biggest benefits of the decanter is its capacity to expel isolated solids from the partitioning zone. This is usually achieved by correlation with:

- Gravity sedimentation: the decanter can accomplish partitions that would be extremely troublesome in a clarifier or lamella separator, also, it produces drier solids.
- Hydrocyclones: the decanter has a significantly higher fluid limit, can deal with considerably higher slurry focuses and makes substantially drier solids.
- Tubular bowl rotators: the decanter offers higher limits, the capacity to deal with think slurries, and persistent operation.

- Imperforate bushel axes: the decanter works persistently, can deal with substantially higher solids fixations, and delivers significantly drier solids.
- Disk stack rotator: the decanter is really nonstop in operation, can deal with significantly higher solids focuses in the nourish slurry and produces drier solids.

In addition, the decanter has a wide variety of potential uses. This combined with its nonstop operation and its accessibility makes it one of the more user friendly options.

Disk Stack Centrifuges

Disk stack centrifuges or disk bowl centrifuges can be used in the separation of larger particle sizes and are ideally used in the separation of a wide range of mixtures containing lower solid concentrations and droplet sizes. They can be used for both liquid-liquid and liquid-solid separations. These centrifuges separate solids and / or liquids using extremely high centrifugal forces. The denserssubstances are forced outwards against the wall of the rotating bowl while the lessedense liquids form the inner layers. The main purpose of the discs is to provide additional surface area which in turn aids in speeding up the separation process. These machines usually have four main zones each with their own function. These include the inlet zone, the disc stack area, the liquid discharge section, the solids discharge section and the separator system. Once the sample is added, the inlet zone accelerates it to the speeds of the rotating bowl. The disk stack area aids in further separation and the effectiveness of the separation is usually dependent on this section. Once separated, the liquid portion must be moved out slowly and without the addition of oxygen or increase in temperatures as this could cause issues in downstream processes. Liquid discharge usually occurs via the application of pressure which results in the pushing of pressure out of the separator via channels in the disks. Solids on the other hand are removed using one of three methods. First is the continuous solids discharge method in which solids and liquids exit via nozzles in the periphery. The second is the intermittent solid discharge methods in which ports are opened in the bowl at regular intervals in which collected solids are removed. The third methods involves manual removal in which the centrifuge itself is stopped and the bowl is opened such that collected solids can then be removed manually (Records and Sutherland, 2001).

Hydrocyclone Centrifuges

Hydrocyclones form a unique design and are one of the simplest centrifuge devices used for separation. Some may say that they are not really centrifuges because while they are still able to separate solid and liquid components they do not have any individual moving parts. They still operate under centrifugal principles with the mixture to be separated being pumped inward at high speed prior to entry. Since the hydocyclone design is such that it forms a conical shape, the flowing liquid subsequently rotates within the cone. The motion of the suspension is induced by theltangential motion of the feed as it is introduced into the system. The centrifugal field is generated by the highly circulating movement of the suspension which creates a core of air on the axis. The larger and heavier solids will migrate to the outside walls of the cylindrical section and then will be pushed downwards to the bottom of the chamber where it is removed. Smaller particles will be dragged upward with the liquids and will be pushed to the top of the cone (Sutherland, 2005). These centrifuges are most often used in the removal of sand or grit from mashed fruit mixtures that require the liquid juice to be separated or in pulp mills where paper fibre is required to be removed.

Pusher Centrifuges

Pusherscentrifuges are commonly used in the chemical and mineral industries. They aid in separation of solid-liquid separations and run on the principle of filtration whereby particles will sediment and form a cake-like build-up on a filter cloth or screen (Figure 50). One of the major advantages associated with pusher centrifuges is the presence of the horizontal pusher which minimises vibrations thereby increasing the stability of the overall centrifuge. The solid and liquid material is also isolated from each other which aids in preventing crosoncontamination. The overall efficiency of pusher centrifuges is dependent on many factors such as particle size, viscosity, solid concentration, cake quality and particle attrition. The particles should be as large as possible because as the size increases, the surface area per unit mass decreases. This in turn means that there is less surface area for moisture to bind to, providing an overall drier cake. The viscosity influences the porosity and subsequent drainage rate with higher viscosities leading to an increase in the resistance experienced by the slurry. This results in a decrease in the overall efficiency of the pusher centrifuge. The third factor is solid concentration, which as the slurry concentration increases, so does the processing rate. The purity and volatile matter present are known to affect the cake quality. Generally a simple washing liquid is passed over the cake to remove any remaining liquid as well as any impurities that may be present. In addition, the amount of volatile matter present is a result of the overall centrifugal force and residence time. With an increasing centrifugal force, the effect of separation in turn increases which ultimately favours more drainage. As the residence time of solisesdecreases, the volatile matters in the discharged cake increases. Lastly particle attrition is also known to affect the efficiency of pusher centrifuges. The centrifugal force is one of the major factors causing attrition of particles which can defined as the progressive loss of material due to collisions and friction (Xu et al, 2014). The overall movement of the pusher plate together with rotation causes the breakage of some particles. Pusugescentrifuges have a wide range of applications in the chemical industry. Many organic intermediates like oxalic acid, nitrocellulose and carboxymethylellulose havc bccn scparated while a range of inorganic compounds like ammonium sulfate, soda ash, sodium bicarbonate, sulfates and sulphides have also been separated.

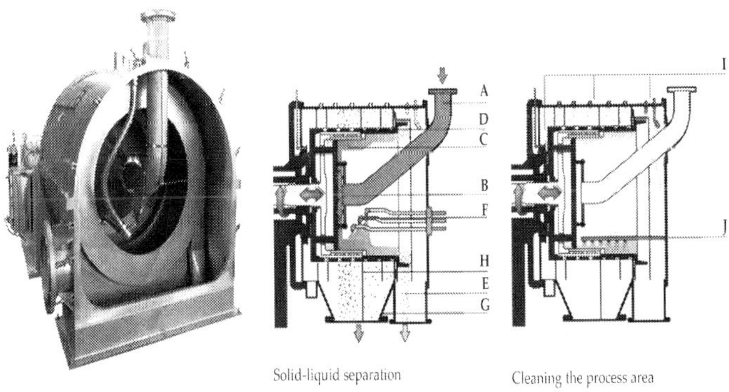

Figure 50: Schematic diagram of a pusher centrifuge indicating the direction of solid and liquid flow. Reproduced with permission from Relex Process Consultancy Ltd.

A) Suspension inlet where the suspension to be separated is continuously fed B) Suspension distributor which spreads the suspension over the entire periphery of the filling area C) First basket stage where the liquid is filtered out into the feed zone resulting in the formation of a stable cake D) Second basket stage which receives the cake from the first basket stage E) Solids discharge where solids leave the centrifuge F) Product washing of any remaining impurities from the initial suspension liquid G) Filtrate housing which collects the filtrate H) Filtrate separation which collects different components of the filtrate which has been separated by separating plates in the filtrate housing I /J) Flushing to aid in preventing formation of crystals and deposition in the solid housing (Centrifuge World, 2017).

Batch Feed Centrifuges

These types of centrifuges are commonly referred to as the first generation of centrifuges machines and are simpler in design and more versatile in their applications. They offer flexible conditions associated with use and allow for adjustments in feed rates and solid concentrations to be made. In addition, they have high separation efficiency and yield high purity of separated products. They can be operated at a range of different speeds and are most commonly used for washing. That being said, they are not widely used in industry as the higher throughput of continuous centrifuges makes the later a more advantageous option. The main disadvantages of these machines however is maintaining stability during loading (Wakeman, 2011). Common examples of batch fugescentrifuges include basket, peeler and pusher-peeler centrifuges. Batch type centrifuges can be broken down into two main designs, namely the top-driven and bottom or under-driven centrifuges.

Batch Type Top Driven Models

These models consist of a rotating basket which is suspended on a vertical shaft. The sides of the basket are usually perforated and covered with a screen on the side while a stationary casing surrounds the basket and collects the filtrate. The motor is place above the entire system and drives the centrifuge from there (Figure 51). Suspensions to be separated are loaded into the basket as it accelerates to maximum speed. The centrifugal force causes the particles to be separated such that they move to the wall of the basket where solids remain on the screen while the liquid passes out through the perforations. Usually there is a discharge present on the bottom of the centrifuge which is raised from where the solids can be cut from the sides of the screen (Majekodunmi, 2015).

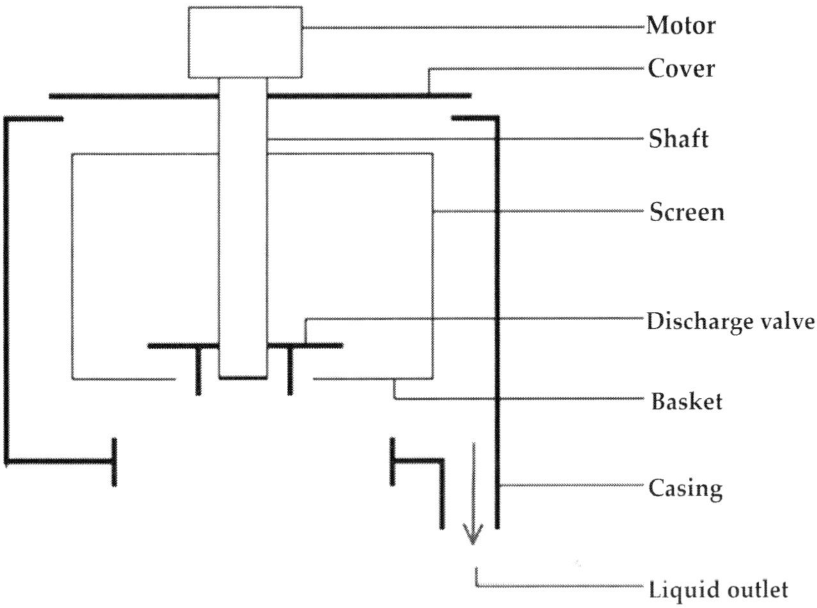

Figure 51: Schematic diagram of a top driven batch centrifuge (redrawn from Majekodunmi, 2015).

Batch Type Bottom Driven Models

These designs consist of a rotating basket placed on a vertical shaft while the motor is situated below. The sides of the basket are usually perforated and covered with a screen on the inside. A stationary case surrounds the basket and this collects the filtrate. The suspension is injected into the basket which subsequently accelerates to maximum speed. The solids are collected on the screen while the liquid passes through (Figure 52). Following deceleration and when the basket is at rest, the cover from the top is raised and any deposited solids are cut out and removed from the screens on the sides.

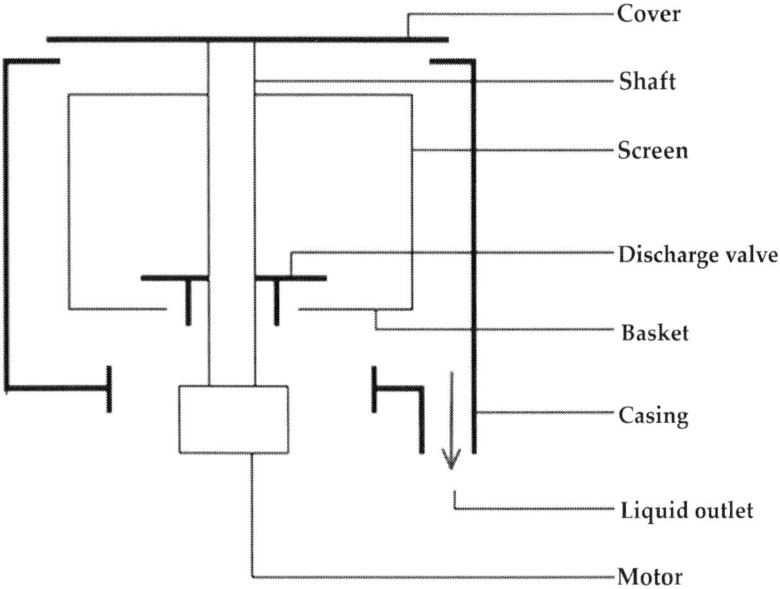

Figure 52: Schematic diagram of a bottom driven batch centrifuge (redrawn from Majekodunmi, 2015).

Tubular Bowl Centrifuges

The tubular bowl design has been used longer than many of the other designs currently available and can be used to separate either solid – liquid mixtures or liquid - liquid mixtures. These separations can be done either manually or automatically with operation being based on a very simple principle. These machines consist of three main parts: a bowl, motor and driver (Figure 53). Generally feed will enter from the bottom of the bowl and incoming mixtures are quickly accelerated to a speed that matches that of the rotor speed. A baffle is then used to separate out the different components with the outer layer consisting of the heavier components and the inner layer consisting of the lighter components. The heavier components which are usually solids will concentrate and attach to the wall of the bowl while the lighter components will float to the surface where it will exit. Solids can only be removed when the machine has decelerated and stopped, followed by scrapping from the screen which may slow down overall processing times. In addition, foaming can also occur so skimming or centripetal pumps should be present in the design (Catalano et al. 2017). Tubular bfgescentrifuges have a wide range of applications in industry. One such application is the purification of industrial oils containing fine amounts of sedimenting solids. They are also used in the food, biochemical and pharmaceutical industries.

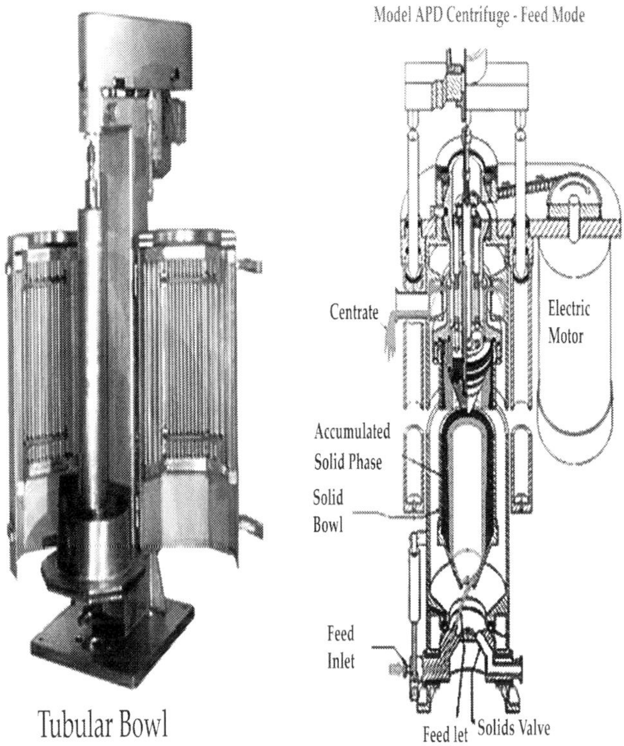

Figure 53: Schematic diagram of a tubular bowl centrifuge. Reproduced with permission from Relex Process Consultancy Ltd.

Solid Bowl Centrifuges

Similarly to the tubular bowl centrifuge, the solid bowl centrifuge is batch operated and does not have the option of continuous removal of sedimented solids. These machines have a large diameter bowl with a removable liner sometimes referred to as a screen (Figure 54). This liner can easily be taken out thereby aiding the removal of solids and cleaning of the centrifuge, ultimately reducing the down time between centrifuge batches. They can be run with a centrifugation speed of 5000 – 10 000 *g* and are most commonly used for collecting expensive precipitates at low concentrations which may require the use of high centrifugal forces for its isolation (Beveridge, 2000). Common modifications of the solid bowl design is the development of a screen bowl centrifuge in which an additional washing step is included followed by enhanced moisture removal to increase the drying of the materials of interest (Andritz, 2017).

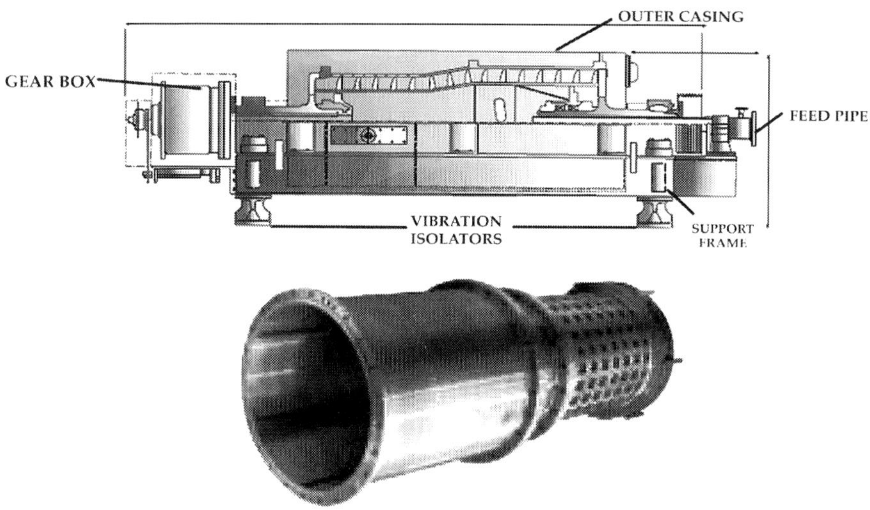

Figure 54: Schematic diagram of a screen bowl centrifuge. Reproduced with permission from Relex Process Consultancy Ltd.

Vertical Basket Centrifuges

These centrifuges utilise a discontinuous filtration process to separate small batches of suspensions. They are commonly used in the pharmaceutical industry for the separation of chemicals. The separated solids are either removed manually from above through the top loader or via an automatic peel mechanism which may use a top loading and bottom discharge method. These centrifuges service a range of industries such as the food, oil and chemical industries. The main advantages associated with vertical basket centrifuges are the low initial investment and ease of operation (Vogel and Todaro, 1996).

Peeler Centrifuges

Peeler centrifuges operate on the main principle of separating solids from liquids by density differences using the centrifugal force. Rotation at high speeds will result in a high centrifugal force which allow suspended solids to settle on the inner surface. One of the atgesadvantages associated with peeler centrifuges is the constant speed which allows all processes like feeding, drying and discharging to be carried out at full uniform speed. Peeler centrifuges are generally used to separate difficult to filter solids from a solid-liquid slurry contaiwneenbetween 5 to 50 % of solids with a particle size ranging from 2 microns to 10 mm. Suspensions to be separated are fed into the basket which is already rotating at full speed. The centrifugal force results in the drainage of liquids from solids through a filter cloth while the solids are collected in cake form in the basket. This may be followed with a simple washing step of the cake to remove any traces of impurities followed by re-drying. A scrapper is then used to peel off the solids which is usually discharged via a chute in the front. These centrifuges are generally used in the food, pharmaceutical, plastic, organic and inorganic chemical industry.

Chamber Bowl Centrifuge

The chamber bowl centrifuge is a tubular bowl that has essential bowl containing tube shaped additions that move the volume of the bowl into a progression of annular chambers, which work in arrangement. Substances enters the focal point of the bowl and the suspension goes through each chamber. The solids settle onto the external mass of each chamber and the cleared up fluid develops as a flood from the biggest measurement chamber. This gadget gives additionally a characterization of the suspended solids: the coarse particles store in the inward chamber and the fine molecules accumulate on the other chambers. The expulsion of sedimented solids requires the ceasing of revolution for manual cleaning.

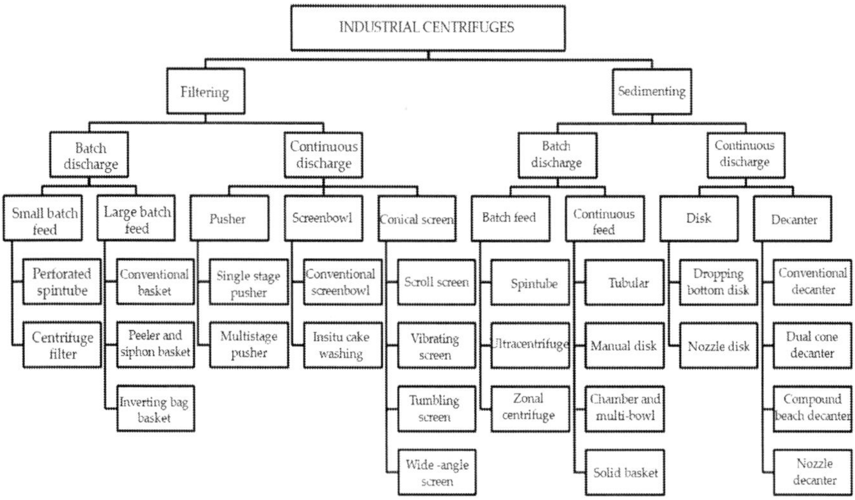

Figure 55: Overview of various industrial centrifuge designs available (Leung, 2007).

CHAPTER 7

INSTRUMENT COMPONENTS

ROTORS

The design of most centrifuges allow for the acceptance of rotors in different sizes, materials and capacities. During centrifugation, particles constantly move along the centrifuge tube distributing the sediment uniformly against the bottom or sides of the tube. After centrifugation is complete and the rotor has ceased turning, the centrifuge tube contains a surface of flat or sloped sediment with a column of liquid above it. The selection of a rotor depends on the sample volume, number of sample particles to be separated, the particle size and density, run time and desired quality of separation. Rotors have three basic designs – swinging bucket or horizontal, fixed angle and vertical or near-vertical rotors.

Fixed Angle Rotors

Fixed-angle rotors are used when rapid sedimentation of small particles is required as they provide faster run times than swinging bucket rotors. These rotors are ideal for differential centrifugation where sedimentation rates differ and runs are used for the separation of biological particles such as nuclei, mitochondria and microsomes. They are the most versatile centrifuge rotors however they are generally used for basic pelleting applications and can be used to pellet particles from a suspension followed by the discarding of excess debris. Fixed angle rotors can also be used for isopycnic centrifugation to pellet bacteria, yeast and other mammalian cells. In this case centrifugation can be continued until the biological particles of interest have reached their isopycnic positions in a gradient. Since the centrifugal field is exerted at an angle, particles will only have to travel a short distance until they reach their isopycnic position in a gradient. These rotors reach much higher speeds of up to 7000 rpm and rcf's of up to 1 000 000 × *g*. Fixed-angle rotors are used when rapid sedimentation of small particles is required. Sample tubes are held at a fixed angle to the vertical with tubes held at a specific angle of approximately 17 to 34 degrees in floor model centrifuges and up to 45 degrees in table top models relative to the vertical axis of rotation (Figure 56). These angles allow the rotor to safely and efficiently achieve the highest possible speed with the shortest possible path length with short run times. During centrifugation, particles move along the side of the tube to form sediment that packs against the side and bottom. The surface of the sediment in this case is parallel to the shaft of the centrifuge. During centrifugation, particles move in a downward spiral manner while migrating towards the wall. The initial sedimenting particles fill the bottom of the tube while additional sedimenting particles collect on the surface of the sediment at an angle. Material collecting on the tube wall is observed only if the initial sample concentration of sedimenting particles in solution is extremely high (Bermann, 1966).

Two major factors will determine the type of fixed angle rotor to be used, namely the desired *g* force or relative centrifugal force and the desired volume. A general rule of thumb is such that the size of the rotor is inversely proportional to its maximum speed. Another important factor when selecting a fixed angle rotor is the k-factor which indicates the pelleting efficiency of the rotor at maximum speed. A low k-factor indicates a higher pelleting efficiency. Thus the k-factor can be useful when comparing speeds at which particles will pellet across a range of rotors. Previous studies conducted by Griffith (1996) showed negligible differences in the total amount of sediment collected between fixed angel rotors utilising different tube angles. An experiment conducted to compare the maximum amount of sediment that can be collected in rotors with an angle of 23 degrees and 34 degrees respectively showed that the concentration of the sample solution was high enough to permit the sediment to reach the centrifuge tube cap. Sample volumes, tube sizes and *g* forces were kept constant for both rotors. Results showed that the rotor with a tube angle of 23 degrees had a total sediment concentration of 23.3g/ml while the rotor with a tube angle of 34 degrees had total sedimentation of 28.2 g/ml. A subsequent centrifugation run with half of the previous amount of sample solution was done yielding results that showed no negligible difference in sediment volumes.

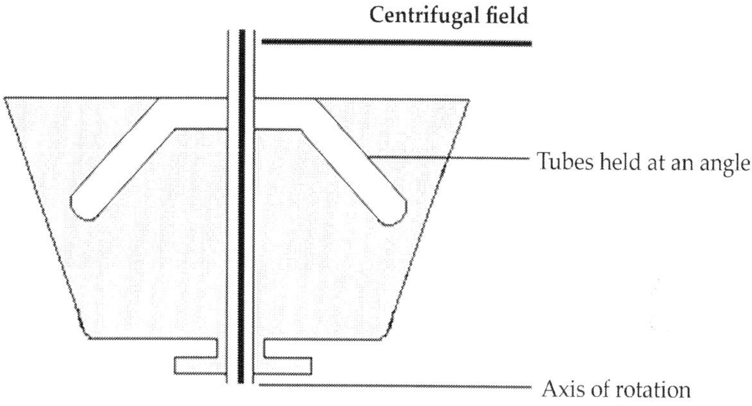

Figure 56: Fixed angle rotor with tubes held at an angle between 14 to 40° (re-drawn with adaptations from Wilson and Walker, 2010).

Tubes and Bottles

Fixed angle rotors can accommodate a variety of tube types such as polyallomer, polycarbonate, stainless steel and so on ; however, the respective fill volumes will vary depending on the kind of tube used. Tubes are usually held at an angle between 20 to 35 degrees to the axis of rotation in each cavity. Specific properties of these tubes are outlined in Table 21 and 22. In addition, some rotors have to be run at a reduced speeds when bottles or tubes are partially filled ; however, the exact run specifications should be followed in accordance with the manufacturers recommendations.

Rotor Preparation and Loading

Most of these rotors have a lid that is secured with a handle while each handle usually has holes so that a metal rod can be inserted to loosen the lid after a centrifugation run is complete. Some rotors may also have small holes to allow any liquid to escape should leakage of samples occur. O-rings may also be found in the rotor lid and these serve to help maintain atmospheric pressure during any subsequent runs. Before use, users should ensure that the metal threads and O-rings are well lubricated and that centrifuge tubes are completely dry on the outside as any moisture that collects between the tube and rotor cavity may lead to imbalances detected during centrifugation runs. All tubes should be arranged symmetrically (Figure) and opposing tubes should be filled to the same level with samples containing the same density.

Safety Checks

If any spacers or adapters are being used they should be placed appropriately. This will depend on the properties of the individual tubes being used. Any cavities without tubes should be left empty. The rotor lid should be placed on firmly and tightened such that the O-rings are fully impressed.

Installing the Rotor

Depending on the required temperature of the run, the rotor an either be refrigerated or warmed beforehand. This will allow for the rotor to reach the required temperature and subsequent equilibrium faster.

To install the rotor, it should be carefully lowered straight down into the centrifuge bowl and onto the drive hub. This step is very important and should be done with care as if it is not done properly, it may lead to rotor misbalancing and improper sedimentation and separation of the required samples.

Sample Recovery from Fixed Angle Rotors

Once the rotor has decelerated and come to a stop, it can be removed usually by lifting it straight up off the drive hub. The handle can be unscrewed and the lid removed. Any spacers should be removed followed by the gentle removal of the centrifuge tubes. Some tubes may have a tube removal tool or forceps can be used.

Swinging Bucket or Horizontal Rotors

Swinging bucket rotors have hanging buckets that hang in a vertical position but which swing out at 90 degrees and upwards to a horizontal position while spinning such that the tube and solution are aligned perpendicularly to the axis of rotation and parallel to the applied centrifugal field (Figure 57). Swinging bucket or horizontal rotors operate to speeds of up to 3000 rpm. The tubes are held in a vertical position when the centrifuge is at rest but move and remain in a horizontal position when the rotor is in motion. Swinging bucket rotors are ideal for separating large volumes of samples up to 12 litres at low speeds. They are commonly used for pelleting studies, rate zonal or isopycnic separations (Voelkl, 2010). Commonly used benchtop centrifuges will use a swinging bucket rotor and are generally used for harvesting whole cells from growth media, high capacity processing of blood collections and large volume tissue culture processing. A swinging bucket rotor system consists of three parts: the rotor body, the buckets and the trunnion pins which hold the bucket containing the centrifuge tubes in place. The rotor body usually attaches to the centrifuge drive and has four or six arms to support the buckets. The buckets are placed onto the arms of the rotor body while the trunnion pins are used to hold the buckets in place. In addition, adapters can be placed into the buckets to hold different tube sizes for larger volume rotors and sealing lids can be attached to provide biocontainment support when runs are conducted with potentially hazardous samples.

The most commonly used method of separation with swinging buckets are rate zonal density gradient centrifugation. This method is commonly employed when maximum resolutions of sample components are required. Since the centrifuge tubes are held horizontally with swinging bucket rotors, the length which particles will need to travel for sufficient separation within the tubes are much longer. Thus sample separation times are also longer. One of the major disadvantages of using swinging bucket rotors is that during centrifugation, the total sample and sedimentation weight is concentrated at the bottom of the bucket. This weight places additional stress on the buckets during centrifugation and can result in reduced efficiency or complete failure of the buckets.

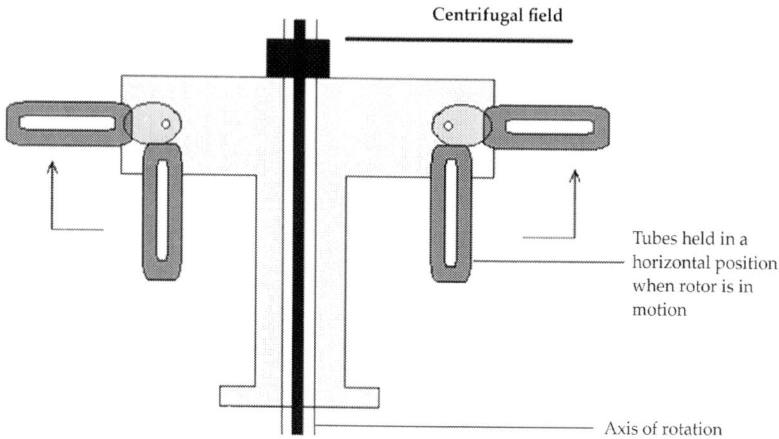

Figure 57: Swinging Bucket Rotor with centrifuge tubes that swing out during rotation (redrawn with adaptations from Wilson and Walker, 2010).

Tubes and Bottles

Tubes in swinging bucket rotors are usually held within the rotor buckets and the buckets itself are attached to the rotor body by either pins or a crossbar. Each bucket is sealed with an O-ring between the bucket and bucket cap to prevent and leakage. Bucket caps are small and should be tightened with a screwdriver.

Rotor Preparation and Loading

Before loading the rotor, buckets and caps should be inspected for any signs of damage or deterioration. When a centrifuge run is being carried out, all buckets must be placed on the rotor body regardless of whether they are loaded with a sample or not. All buckets should be inspected for the presence of O-rings. If these are not present then the rotor should not be operated as the bucket contents may be lost leading to rotor imbalance and possible failure of the rotor. When loading the rotor, loaded buckets must be arranged symmetrically on the rotor while any opposing tubes should be filled to the same level with liquids of the same density.

Safety Checks

Centrifuge tubes should be filled up to 3 mm from the top and should be balanced accordingly. The difference of any two tubes arranged symmetrically should be less than 0.2 g. tubes should be filled with an injector such that a small quantity of bubbles remain.

Installing the Rotor

If runs are being carried out at temperatures other than room temperature, the rotor can be warmed or refrigerated for faster temperature equilibrium. The location of drive pins should be identified prior to loading of the rotor into the centrifuge bowl. This can be done by identifying the two small rotor indentations of the rotor adapter. The rotor should be lifted with both hands and lowered down straight onto the drive hub. Care should be taken not to lift the rotor near the adapters as the buckets may be dislodged and lead to imbalances in the rotor and sample from centrifuge tubes spilling.

Sample Recovery from Swinging Bucket Rotors

The rotor should firstly be gently removed from the centrifuge bowl by lifting it vertically and off the drive hub. The rotor should then be placed on a flat counter or the respective rotor stand. The buckets should be carefully removed by lifting and unscrewing the hinge pins. Any spacers or adapters should be removed too. If there is evidence of leakage after or during disassembly, then the appropriate decontamination procedures should be applied.

Vertical Rotors and Near Vertical Rotors

Vertical tube rotors are also known as zero angle fixed rotors and have the centrifuge tubes aligned vertically in the body of the rotors at all times (Figure 58).They are most commonly used during ultracentrifugation for isopycnic separations such as the banding of DNA in a caesium chloride solution or to separate plasmid DNA, RNA and lipoproteins. These rotors are known to have the shortest path length and radial distance which in turn results in shorter run times. Unlike fixed angle and swinging bucket rotors, sedimentation occurs across the diameter of the tube (Griffith, 2010).

Tubes and Bottles

The tubes in vertical tube rotors are held parallel to the axis of rotation and usually have plugs that are screwed into the rotor cavities. These plugs serve to maintain the position of the tubes in the cavities and provide support any form of hydrostatic forces. Tubes in near vertical tube rotors are held at an angle of approximately 7 to 10 degrees with respect to the axis of rotation. This angle is known to aid in reducing run times and the subsequent separation of those particles that may not necessarily band out. During centrifugation, sedimentation occurs across the diameter of the tube instead of down the length of the tube as it would normally occur in fixed angle rotors or horizontally as in swinging bucket rotors. One of the major disadvantages of this rotor is that resolution is lost when samples with multiple particles are required to be separated as collected sediments will remain on the walls of the centrifuge tube until the run is terminated. After the rotor decelerates and stops, any sediment that is pelleted during centrifugation usually slides off the tube walls and re-contaminates the separated sample zones that may be present in solution.

Rotor Preparation and Loading

Vertical rotors are not suitable for pelleting of samples but are most efficient for isopycnic separations. This is because samples are not separated down the length of the centrifuge tube but rather across the diameter of the tube resulting in significantly shorter run times as when compared to swinging bucket rotors. Vertical rotors may be subdivided into true vertical rotors and near-vertical rotors. Vertical rotors contain sealed centrifuge tubes that are held parallel to the axis of rotation and which are restrained in the rotor cavities by screws, special washers and plugs. They are specialised with their most popular use being during ultracentrifugation for isopycnic banding of DNA in caesium chloride. Near-vertical rotors have a reduced tube angle of 7 to 10 degrees which results in much shorter run times as compared to fixed-angle rotors and are useful for gradient centrifugation of biological particles.

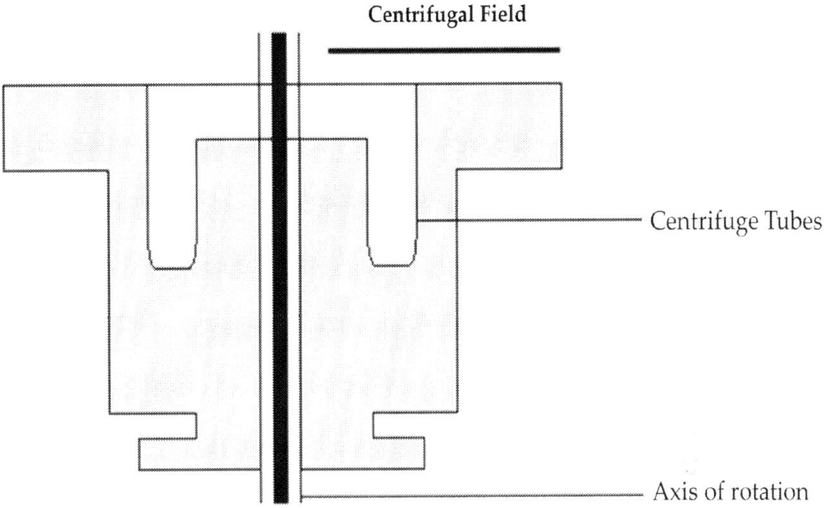

Figure 58: Vertical Rotors with centrifuge tubes set at 0 degrees from the axis of rotation (redrawn with adaptations from Wilson and Walker, 2010).

Safety Checks

The rotor, plugs and all other components should initially be inspected for any signs of damage or corrosion as these rotors can reach high speeds and this may cause any damaged components to fail. There should also be no condensation within the rotor cavities as this could cause future imbalances during centrifugation runs. The chemical compatibilities of buffers and samples with centrifuge tubes should be checked beforehand.

Installing the Rotor

The rotor should be installed carefully and lowered slowly down vertically onto the drive hub. This should be done slowly to ensure that loaded samples are not disturbed and the imbalance detector is not alerted.

Sample Recovery from Vertical and near Vertical Rotors

Like others, the rotor should firstly be gently removed from the centrifuge bowl by lifting it vertically off the drive hub. Any rotor plugs or spaces should then be removed using the appropriate tools. If there is evidence of leakage after or during disassembly, then the appropriate decontamination procedures should be applied.

MATERIALS USED FOR ROTORS

The choice of material used to manufacture rotors will depend on the exact type and application of each one (Table 17). High speed ultracentrifuge rotors are usually made out of titanium and aluminium alloys while low speed centrifuges can be made of bronze, aluminium or plastic materials such as polypropylene. Since these rotors reach such high speeds, having a uniform surface is particularly important. This is usually achieved by a finishing step during manufacturing whereby the rotors are coated with an anodized or varnish coating such as an aluminium alloy.

Aluminium Rotors

Aluminium rotors are relatively light and easy to handle but must be treated with care because this material is less durable and particularly susceptible to corrosion. The thin film of oxide protecting the surface is very brittle and tends to crack during centrifugation giving aggressive chemicals access to the reactive aluminium alloy. Corrosion is usually caused by acids, alkalis and high concentrations of salt solutions. Even dilute solutions of salts contained in buffers can become concentrated as a result of evaporation and lead to corrosion of rotors. Aluminium rotors are usually anodized to decrease corrosion and further wear and tear but this is not always the most effective method for protection of the metal. If the anodizing film is damaged in any way during centrifugation, any liquid present in the tube cavity of the rotor will be forced into the crystalline structure of the metal which eventually leads to stress corrosion. Stress corrosion is the main cause of failure in aluminium rotors.

Titanium Rotors

Titanium rotors are generally stronger than other rotors and much more resistant to corrosion because the titanium alloy is not affected by most acids, alkalis or salt solutions. This makes them the preferred rotor for rapid separation runs at high speeds or when corrosive chemicals are used. They are also exposed to less fatigue, are lighter in weight and have a higher strength thereby allowing for faster acceleration and deceleration as when compared to other metal rotors. Titanium rotors can also withstand more stress cycles when run at full speed as when compared to aluminium rotors however, repeated run cycles of acceleration and deceleration from maximum speed will eventually weaken the rotor (Voelkl, 2010).

Carbon Fibre Rotors

Carbon fibre rotors are extremely light, up to 60% lighter than titanium rotors and 50% lighter than aluminium rotors. This makes them easier to load into a centrifuge as well as carry and clean. In addition, many are able to reach maximum speeds faster than metallic rotors. Carbon fibre rotors also have the highest strength to weight ratio of all rotor materials making the lifespan of carbon fibre rotors greater than other metal rotors. Unlike aluminium and titanium materials which are prone to corrosion, carbon fibre is corrosion-resistant. As a result these rotors are able to withstand exposure to common corrosive liquids such as most organic solutions. They can also be used with most alkaline laboratory detergents and commercially available alkaline solutions for radioactive decontamination.

Table 17: Comparison of rotor material characteristics

Aluminium	Titanium	Carbon Fibre
Highly susceptible to acid, alkali and salts corrosion	Moderately susceptible to acid, alkali and salt corrosion	Not susceptible to corrosion
Anodizing may lead to stress corrosion	Repeated run cycles may weaken the rotor	Longer life span due to high strength to weight ratio and no stretching or elongation during centrifugation
Made with a heavy material that requires more energy to rotate	Made with heavy material that requires more energy to rotate	Lightweight material that insulate heats and maintains sample temperature. Material is environmentally friendly

Selection of Rotors for Centrifugal Separations

When choosing a rotor for centrifugation experiments, a range of factors need to be taken into consideration (Table 18). The selection of a rotor depends on the type of samples to be centrifuged, the sample volume, particle size and density, the concentration of particles to be separated, the run time and the quality of separation required.

Table 18: Summary of common centrifugal separation methods and related features

ROTOR TYPE	SUMMARY OF FEATURES
Fixed-angle rotor	Most commonly used type of rotor Tube holes are angled between 0 – 30 ° Ideally suited for pelleting
Vertical Rotor	Centrifuge tubes are set in a vertical position Shorter sedimentation distances enable faster separation Suited for density gradient centrifugal separations
Swinging Bucket	Centrifuge tubes are maintained in a horizontal position during centrifugation Longer sedimentation distances which result in optimized separation accuracy Suited for sucrose density gradient centrifugation separation.

For these reasons, rotor properties such as the type of material it is made of, the geometry and capacity as well as the maximum speed should be taken into consideration (Table 19). Swinging bucket rotors allows the centrifuge tubes to hang on to the hinges so that tubes will reorient to the horizontal as the rotor accelerates. Fixed-angle rotors are made of a single block of metal and hold the centrifuge tubes at a predetermined angle while zonal rotors are designed to contain a large volume of sample in a single central cavity rather than in tubes.

Table 19: Types of Rotors and their respective applications

Type of rotor	Pelleting	Rate-Zonal	Isopycnic
Fixed angle	Excellent	Limited	Variable*
Swinging Bucket	Inefficient	Good	Good**
Vertical	Not suitable	Good	Excellent
Zonal	Not suitable	Excellent	Good

*Good for macromolecules, poor for cells and organelles

**Good for cells and organelles, precautions should be taken when used with CsCl

The separation of particles will vary within the three main rotor types (Table 20). Orientation of samples within the three main types of rotors used in routine biochemical centrifugation runs are shown below with a cross sectional diagram of a centrifuge tube position within the different rotors, namely fixed angle rotors, vertical tube rotors and swinging bucket rotors. The diagrams illustrate the movement of biological samples when centrifuge tubes are maintained at rest, during the main centrifugal separation phases and after the centrifugation run.

Table 20: Orientation of samples in commonly used rotor types

	At rest	At speed	After run
Swinging bucket		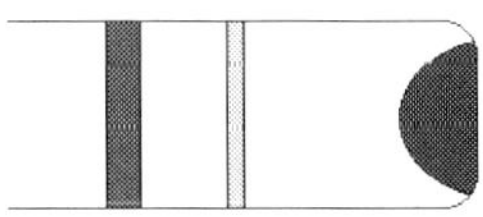	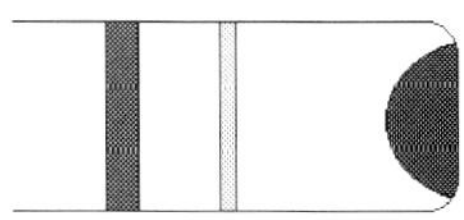

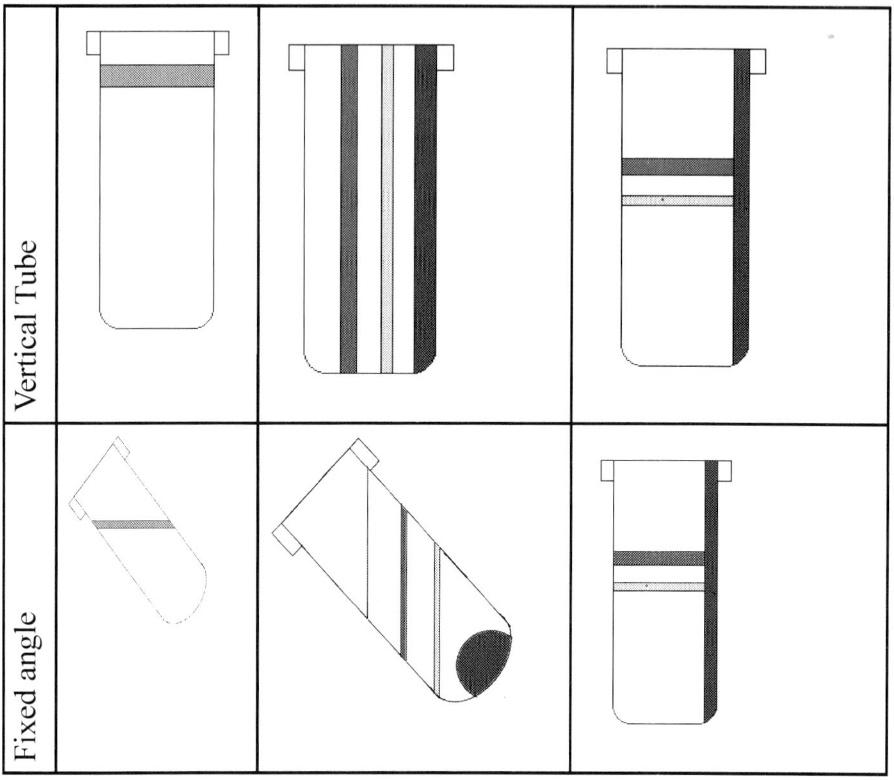

(redrawn and adapted from Griffith, 2010)

Allowable Rotor Speed

Rotors should only be run at the allowed speed as specified by the manufacter and should never be operated higher. Rotor speed is limited by four main parameters. These include 1) the allowable speed based on the density of the sample 2) the allowable speed based on the density gradient medium 3) the allowable speed based on the combination of parts used and, 4) the allowable speed based on the life of the rotor (Hitachi 2013).

Calculating the allowable Rotor Speed based on the Density of the Sample

When centrifuging a sample with an average density of greater than 1.2 g/m and if the required speed is unknown it can be determined using the following equation:

Allowable speed (rpm) = maximum speed (rpm) × √(1.2 / average density of the sample in grams per ml)

If samples to centrifuged are placed in seal tubes and are to be centrifuged in neo-angle or vertical rotors, one of two options are available:

- If the sample has an average density of less than 1.7 g/ml then the above rotors can be used at their maximum speed
- If the sample has an aveage density that is greater than 1.7 g/ml then the following equation can be used:

Allowable speed (rpm) = maximum speed (rpm) × √(1.7 / average density of the sample in grams per ml)

Allowable Speed based on the Density Gradient Medium

Cesium chloride is a commonly used density gradient medium for separation of various particles. If a CsCl solution with a high density is used, some form of precipitation may occur depending on the rotating conditions. This precipitation will result in the formation of CsCl crystals which in turn exert a heavy load on the rotor. This crystallization may also affect the separation properties of the sample hence the need to use a solution that will not crystallize with the allowed rotor speed.

Allowable Speed based on Centrifuge Parts

The respective centrifuge parts like, tubes, bottles, caps, adaptors and rotors to be used will all influence the allowed speed based on the individual properties of each. It is best to consult with the manuel supplied from the manufacturer to determine what is an acceptable speed that can be used with the chosen combination of adapter, tube and rotor.

Allowable Speed based on the Life of Rotors

All rotors will have specifications as to the initial maximum speed followed by the changes in maximum speed after being used for a certain period of time. For example, the speed of aluminium rotors should be limited after 1000 runs or 2500 hours of use. In addition, any potential signs of damage to the rotor may also require a reduction in the maxiumum speed and this can be determined from the manufacturers manuel.

Care of Rotors

Proper rotor care is essential for ensuring safety and longevity of both rotors and the centrifuge itself. Regular maintenance of rotors is imperative to reducing the risk of accidents which can in turn save users time and money while also prolonging the rotor life span. All rotor manufacturers will usually provide guidelines regarding acceptable cleaning methods and disinfection agents so it is highly recommended that users read the given instruction manuals before use.

Handling of Rotors

Rotors should always be handled with care since the protective anodised coating is very thin and improper handling and usage will lead to scratching and corrosion. They should be stored in a clean dry environment away from any moisture.

Cleaning of Rotors

Since different rotors are made from different materials the disinfection products used for cleaning will vary depending on the respective materials. It is important to select the appropriate cleaning agent for specific rotor materials in order to preserve and prolong rotor lifespans. For example bleach should not be used on aluminium rotors as it will aid corrosion but it can however be used on carbon fibre rotors. Rotors should be thoroughly washed with deionised water and allowed to dry upside down in a warm place. This must always be done since moisture can also serve as a source of corrosion. Swinging bucket rotors should never be completely immersed in water as the bucket hanging system is difficult to dry.

A simple set of general cleaning steps are outlined below:

- Begin by switching off the centrifuge and removing any centrifuge tubes, buckets and the rotor from the centrifuge.
- For refrigerated centrifuges, leave the machine lid open so that any ice in the rotor chamber surface can be defrosted.
- If the centrifuge has a water collection tray, then remove this once the ice has defrosted and empty any collected water.

- Wipe down the collection tray and rotor chamber with a moist cloth.
- Wipe down the outside of the centrifuge with a moist cloth and a neutral detergent.
- If you are unsure of what samples have been placed in the machine during previous runs then a laboratory coat, gloves and goggles should be worn when cleaning.
- Rotors should be thoroughly dried and stored upside down to ensure all moisture is removed
- After cleaning, a small amount of centrifuge lubricant such as glycerol should be applied to bucket grooves, pivots and rubber seals.

Rinse the Rotor

Regardless of the material the rotor is made with, every rotor should be rinsed with water after each use to remove any residual samples. This should subsequently be followed by drying to ensure all moisture is removed. If any debris is still present and cannot be removed by simple rinsing, then a mild detergent with a soft cloth or gentle brush can be used to remove it. Metal cleaning agents like steel wool, wire brushes, abrasives or even sandpaper should not be used as they may scratch or damage the anodised rotor coating and increase the risk of corrosion (Schroeder, 2013).

Disinfection of the Rotor

If centrifugation of potentially infectious substances is carried out, then a disinfection step should be done after the run is completed. Different rotors will require different disinfection productretherefore, it is important to select the appropriate cleaning agent depending on the rotor material. As a general rule of thumb, acetone, caustic detergents or detergents that contain chlorite ions should not be used as this may lead to corrosion.

Sterilizing of Rotors

In certain instances, rotors may be required to undergo sterilization. This is most often done by autoclaving or through UV exposure to ensure complete removal of any infectious agents. This process will generally remove all forms of microbial life including bacteria, viruses, fungi and spores. Common autoclaving methods for heat sterilization will be carried out at 121 °C, at 2 bar atmospheric pressure for 15 to 20 minutes. All rotors, regardless of the material it is made out of are autoclavable which means this is one of the easiest and most commonly used methods of sterilization (Lawson et al., 2007).

COMMON CAUSES OF ROTOR FAILURE

Centrifuges can last for up to 15 years and overtime will have multiple userserHowever, with time and repeated use, rotors age and become fatigued. Thus it is important to maintain rotors properly and be aware of the factors that can lead to rotor failure. These include stress, fatigue, stress corrosion and corrosion.

Stress

High rotational speeds generate an increased stress on rotors which may in turn cause it to change in size. Depending on the amount of stress, the rotor will stretch but can subsequently return to its original dimensions after the load is removed. However, at a certain level of stress, the elastic limit will be exceeded such that the rotor will not regain its original size and shape. This damage may lead to complete failure of the rotor. Paying attention to instructions provided by manufacturer's regarding run cycles will help in preventing unnecessary stress on the rotor. Common causes of stress include running above the recommended maximum speed, overuse and age, using the incorrect tubes and precipitation and moisture inside the cavities of the rotor.

Fatigue

After an extended period of use and stress cycles, a rotor will eventually suffer fatigue. Repeated runs up to maximum operating speed followed by rapid deceleration will result in the continuous stretching and relaxing of the metal which in turn leads to changes in the microstructure of the metal. Over time, these changes in the microstructure will lead to the formation of small cracks which will inevitably become larger and lead to complete failure of the rotor.

Stress Corrosion

Corrosion is caused by a range of factors such as moisture, detergent chemicals and alkaline solutions as well as sample materials. One of the most common after effects of corrosion is pitting. This leads to a reduction in the amount of metal available to bear any kind of stress during centrifugation runs. When there is a concentration of stress in a pitted area, corrosion and cracks will start to develop. This combination of stress and corrosion is known as stress corrosion and will lead to failure of the rotor much sooner.

Corrosion

Rotors are generally chosen for their mechanical properties. While titanium rotors are corrosion resistant, aluminium rotors are not. When corrosion occurs, the metal of the rotor is weakened and less able to bear the stress from subsequent centrifugation runs. This will in turn cause the rotor to fail quickly and at lower stress levels than in an uncorroded rotor. A range of factors can lead to corrosion such as improper rotor usage and use of incorrect chemicals and detergents.

ROTOR MANAGEMENT

Many modern centrifuges have built-in features to aid centrifugation and rotor management regardless of the age of the machine or the level of the user. Rotor management allows users to manage and maintain the centrifuge and rotors by monitoring the rotor serial number, total number of hours used and total number of cycles that have been completed. By taking these factors into consideration, the centrifuge itself can alert the user of the appropriate time to retire a rotor or to call a technician in for routine maintenance. In addition, in some modern centrifuges the serial number of rotors will indicate the age and number of runs done. Thus when it has exceeded the maximum number of runs, the centrifuge will not allow the rotor to be used at certain speeds. In addition, each rotor has a specific identification code (ID) that is registered with specific centrifuges. This ID prevents nonspecific rotors from being run in certain centrifuges. Rotors are usually identified in a variety of ways. Some rotors are identified by a magnetic configurations on the bottom of the rotor while others can be identified via inertia readings that are taken during acceleration. These readings compare the amount of energy being used to turn the rotor by the motor to the on-board database. This database is usually programmed with the correct amount of energy required for all rotors usable within that unit. In addition, the on board system will also check and confirm that the programmed speed does not exceed the maximum rpm for that specified rotor. These rotor inspections can be routinely conducted by laboratory technicians who can subsequently educate laboratory staff on the warning signs preceding rotor failure.

Simple steps to improve Rotor Management

As a general rule of thumb all users of the centrifuge and rotor should be well trained in basic rotor management steps. The centrifuge and rotor manual should always be read before the rotor is used for the first time as well as if there are any uncertainties or discrepancies. All manufacturing specifications should be incorporated into operating procedures. Attention should be paid to the maximum speed and sample density ratings stated by the manufacturer for each rotor. This is important to prevent stress and subsequent rotor failure. Ensuring the correct rotor, tubes and adapters are used with the correct centrifuge is imperative. In addition, the purchase date of each rotor should be recorded together with the manufacturing date, serial number, warranty period and retirement recommendations. Adequate storage conditions should carried out in a clean, dry environment. As a general rule of thumb, fixed angle vertical tube and near-vertical tube rotors should be stored upside down, with the lids removed while swinging bucket rotors should be stored with the caps of buckets removed.

Preventing Corrosion

Corrosion can be prevented by following a few simple steps as outlined below:

- Ensure rotors are routinely inspected by trained personnel.
- All centrifuge components such as rotors, tubes and caps should be examined for signs of corrosion, pitting or cracks before each run.
- Water should not be dispensed directly into buckets when balancing.

- Buckets and rotors should be washed immediately after centrifugation runs to ensure any corrosive substances are removed.
- Users should be mindful of the rotor material and care should be taken when handling such that they are not scratched.
- Adaptors should be disabled and removed immediately after use.
- Rotors should be wiped using non-abrasive cloths and left to air dry.
- Rotor protection spray and lubricants should be regularly applied.
- Rotors and tubes should be stored upside down and with the lids off.
- O - Rings and centrifuge parts should be greased regularly in accordance with manufacturer's recommendations. In addition, they can also be greased at least once a week for machines that are in regular use or before each run for those used occasionally.

Containment of Aerosols during Centrifugation

Aerosols can be contained by following a few simple steps as outlined below:

- Hazardous substances should only be centrifuged in tubes that are capped and leak-proof.
- If the samples to be centrifuged contain pathogenic components then these tubes should only be opened in a Microbiological Safety Cabinet Class I or II depending on the type of biological agent present.

ADDITIONAL CENTRIFUGE COMPONENTS

Motor

The motor of the centrifuge is of utmost importance as it is the type and quality of the motor that will determine the life span and effectiveness of the centrifuge machine as a whole. Generally most centrifuges have high torque motors. Generally series wound DC motors are used which rotate faster as the voltage increaserllyhowever, occasionally AC motors are also used (The Biomedical Scientist, 2013). Both DC and AC motors are high speed and employ direct drive systems. Control of motors has been achieved by controlling the current that reaches the induction motor which will in turn drive the shaft and rotor.

Imbalance Detectors

Most centrifuges are equipped with an internal imbalance detector that monitors the rotor during centrifugation runs. It will result in automatic shutdown should the rotor be imbalanced. Some rotor imbalance detectors contain a moveable rod that is thrown out should a severe imbalance be detected. This rod is usually mounted beneath the rotor and causes the motor driving the rotor to be de-energised thereby stopping the rotor from running.

Tachometers

A tachometer measures the rotational speed of the motor. Modern centrifuges use electronic tachometers in which a magnet rotates around a coil to produce a current that can then be measured.

Safety Lids

All centrifuges have a safety lid that requires that seals the machine shut before use. Most modern centrifuges have a door locking mechanism to prevent the lid from being opened while the instrument is running. This door locking mechanism can be manually released should there be a power failure or if the safety latch fails to manually release the door allowing for samples to be retrieved.

Refrigerators

Refrigeration constitutes an important component of centrifuges as a great deal of heat is produced when they rotate at high speeds. The main purpose of refrigeration is to cool the chamber and rotor down so that temperature sensitive samples will not be affected during centrifugation runs. Some centrifuges even enable the rotor and chamber to be precooled before a centrifuge run begins in an attempt to maintain temperature control.

Braking Systems

Braking systems within a centrifuge allow for rapid rotor deceleration. This occurs by reversing the polarity of the electrical current to the motor. Older machines may still have a mechanical brake for stopping the centrifuge run.

SAMPLE CONTAINERS

Centrifuge tubes and bottles are available in a range of different sizes and materials such as glass, polypropylene, polycarbonate, stainless steel and nylon. The type of containers used will depend upon the nature and volume of the sample to be centrifuged along with centrifugal forces that will be applied (Table 21). While no single tube, bottle design or material meets all the required applications, the use of specific bottles or tubes should be based on a range of factors like the centrifugation technique to be used, rotor being used, samples to be separated and related volume as well as band visibility. In addition the chemical properties of the sample, solvent and gradient medium should be taken into consideration as well as the temperature and speed (Table 22). Glass centrifuge tubes are suitable only for centrifugation at low speeds. Thin walled tubes may be used in swinging bucket rotors because of the protection provided by the surrounding bucket while thick walled tubes are required with fixed angle and vertical tube rotors. The centrifuge tubes should be filled to the appropriate level and they should be capped or bottled depending on the speed and type of container used. Tubes should be autoclaved only if absolutely necessary and at 121°C for 15 min. Cleaning of plastic tubes should be avoided in automated dishwashers or glassware washers as excessively hot temperatures may affect the effectiveness of the tubes.

Table 21: Types of centrifuge tubes and rotor compatibility

	Rotor Type		
Tube Type	Fixed angle	Swinging Bucket	Vertical
Thin walled open top	N	Y	N
Thick walled open top	Y	Y	N
Thin walled sealed	Y	Some	Y
Oak ridge	Y	N	N

N = no, Y = yes, S = some

Chemical Compatibility of Centrifuge Tubes with Solvents, Samples and Gradient Media

Centrifuge tubes or bottles can be made from a range of materials, each of which has varying chemical properties and centrifugation run requirements (Table 22).The chemical compatibility of centrifuge tubes and bottles with the sample being separated is an important factor to consider. While sucrose and many salt solutions are neutral and can be used in almost all kinds of tubes, more alkaline solutions cannot be readily used in every kind of tube. This is especially the case for polycarbonate tubes which are known to be incompatible with Dimethyl sulfoxide (DMSO) and other solutions. In addition, tubes should be chosen with gradient formation, fractionation and subsequent downstream processes in mind. Some materials like polyallomers are not as translucent as others which may in turn make indistinct bands not as visible. If band visualization is important then polycarbonate or cellulose propionate tubes should be used as these are more transparent. In addition, if fractionation collection methods involve the puncturing or slicing of centrifuge tubes, then thin, flexible tubes should be used.

Table 22: Overview of centrifuge tube and bottle materials and their respective properties

Material	**Properties**
Polycarbonate (PC)	Strong, transparent and can be autoclaved at 121 °C or less. These tubes or bottles should be used at temperatures between 4 to 25 °C.
Polyethylene terephthalate (PET)	Transparent and cannot be autoclaved. This material is weak and easily can be easily punctured.
Polyethylene (PE)	This material is opaque and cannot be autoclaved. It is resistant to most chemicals and should be used with temperatures between 4 to 20 °C.
Polypropylene copolymer (PA)	Strong and should be autoclaved at 115 °C or less. These tubes or bottles should be used with a temperature of between 4 to 25 °C.
Stainless Steel (SS)	This material is excellent in providing chemical and heat resistance. It is usually autoclaved at 121 °C of less.

Thin-Walled Centrifuge Tubes

These tubes are usually used in angle and vertical rotors and should be used with caps. They should also be filled to the top while use in swing rotors require tubes to be filled with a headspace of between 2 to 3 mm from the top and can be used without caps. Thin-walled open top tubes are usually used in swinging bucket and fixed angle rotors. These thin-wall tubes are designed for one-time use only and should be discarded appropriately after use.

Thick-Walled Tubes

Thick-walled tubes are reusable and offer the option of centrifuging with partially filled tubes with or without caps. They can be used in angle and swing rotors. There are no volume restrictions when using these tubes in angle rotors however; they should be filled up to the top, leaving a headspace of approximately 3 mm from the top in swing rotors. Most often, tubes without caps are partially filled however the tube angle will determine the maximum volume that can be placed within the tube. If greater volumes are required to be filled then tubes with caps should be used.

Centrifuge Bottles

These consist of a larger container with a screw cap and can be made out of a range of materials such as polycarbonate, polypropylene and polyallomer. They may be hard and clear or translucent depending on the material from which it is made which in turn makes it easier to identify the solutions within. Centrifuge bottles are larger than many of the other centrifuge designs. They have wider mouths making it easier to fill and decant materials. In addition, many of these bottles will have gradations on the side to assist with measuring volumes and sediment concentrations.

Polyallomer Tubes

These tubes are made of a copolymer consisting of ethylene and propylene and can be either transparent or translucent in appearance ; however, this will depend on the wall thickness. These tubes are usually nonwettable although some polyallomer tubes can be chemically treated to make them more wettable. They can be used with a range of gradient media and detergents regardless of whether they are acidic or basic. They also perform well with many alcohol solutions, DMSO and even some organic solvents. There are two main varieties of polyallomer tubes namely the thin walled open top tubes and the thick walled open top tubes.

Polycarbonate Tubes

Polycarbonate tubes have a glass-like appearance and are tough, rigid and nonwettable. They can be used with or without caps in fixed angle and swinging bucket rotors. Sterilization can be done via autoclaving however this may greatly reduce the lifespan of the tubes. In addition, washing with alkaline detergents may cause tube failure and subsequent cracking which will in turn affect the overall performance of particle separations. Over time, the cracks may increase in size and visibility and may lead to the leaking of samples. These tubes are known to have a good tolerance to many kinds of gradient media except for those with pH 8 or greater (alkaline). They can be used with weak acids but not with any bases, alcohols or organic solvents.

Polypropylene Tubes

Polypropylene tubes are translucent and can be used with a variety of gradient media including alkaline solutions. They can also be used with acids, bases and alcohols although the best results may not always be obtained while unsatisfactory results are usually obtained when used with organic solvents. These tubes are reusable and can be used with or without caps in fixed angle rotors.

Polyethylene Tubes

Polyethylene tubes can be transparent or translucent and are reusable. They can be used with strong acids and bases however they cannot be autoclaved. They are usually used without caps in swinging bucket rotors and with or without caps in fixed angle rotors.

Cellulose Propionate Tubes

These transparent tubes are usually designed for one-time use and should be used without caps during centrifugation. Tubes should be full when centrifuging and are generally used without caps. They can withstand a range of gradient material including alkaline solutions ; however, they do not work well with acids or alcohols.

Stainless Steel Tubes

These tubes are much stronger and provide increased resistance to a range of gradients and heat although they cannot withstand most acidic and basic solutions. They can be centrifuged when filled to any level as long as the tubes are balanced. Despite being stronger, these tubes are also much heavier and will have to be run at reduced speeds. One of the major advantages with these tubes are that if they are maintained properly, they will have a very long lifespan. Sterilization usually involves autoclaving followed by proper drying before storage.

Table 23: Overview of centrifuge tubes and bottles properties

	Thin wall polyallomer	Thick wall polyallomer	Polycarbonate	Polypropylene	Polyethylene	Cellulose propionate	Stainless steel
Optical property	Transparent	Translucent	transparent	Translucent / transparent	Transparent / translucent	Transparent	Opaque
Puncturable	Y	N	N	N	Y	N	N
Sliceable	Y	N*	N	N*	N	N*	N
Reusable	N	Y	Y	Y	Y	N	Y
Acids (dilute and weak)	S	S	M	S	S	S	S
Acids (strong)	U	S	U	S	S	U	U
Alcohols (aliphatic)	U	S	U	S	S	U	S
Aldehydes	M	M	M	M	S	U	S
Bases	S	S	U	S	S	U	M
Esters	U	M	U	M	S	M	S
Hydrocarbons (aliphatic)	U	M	U	S		S	S
Hydrocarbons (aromatic and halogenated)	U	U	U	M	U	S	S
Ketones	U	M	U	M	M	U	M
Oxidising agents (strong)	U	U	M	M	M	M	S
Salts	S	S	M	S	M	S	M

Y = resistant, N = not resistant, S = satisfactory resistance, M = marginal resistance, U = unsatisfactory resistance (adapted from Beckman, 2007)

Adapters

The purpose of adapters is to aid accommodation of different centrifuge sizes. These adapters line together the centrifuge tube with either the tube cavity or rotor bucket and are generally made from white polyacetal material (Figure 59). Tubes used with adapters can be filled and capped depending on the type of tube and the design of rotor being used. In addition, a reduction in speed may be required due to the added density of the adapter.

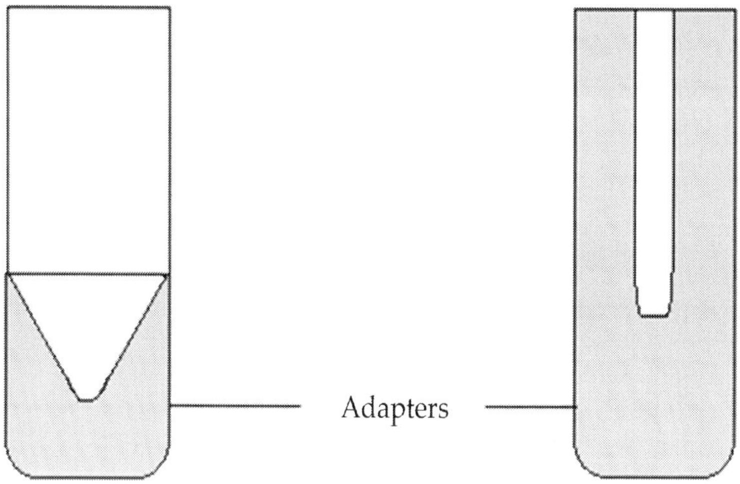

Figure 59: Different adapters fitted on to centrifuge tubes (redrawn from Beckman Coulter, 2007).

Spacers

Certain centrifuge tubes require spacers to support the top of the tubes during centrifugation. Depending on the size of the tube and rotor being used, a floating spacer or a combination of different spacers may be used (Figure 60). For example, in swinging bucket and fixed angle rotors, the top of the tube must be supported while in near vertical and vertical tube rotors, the entire tube cavity must be filled. Most spacers can withstand temperatures between 2 and 25 °C. Centrifugation above 25 °C will lead to deformation of the spacer and subsequently deformation of the centrifuge tube holding the spacer.

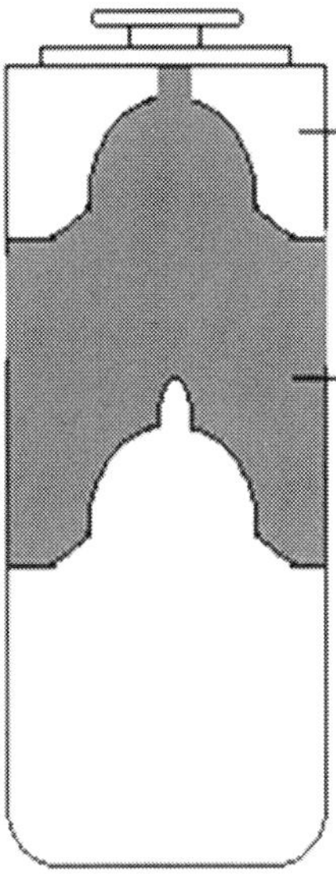

Figure 60: Centrifuge tube with a floating spacer fitted (redrawn from Beckman Coulter, 2007).

Washing of Centrifuge Tubes and Bottles

There are a range of different washing procedures available ; however, centrifuge users should always consult with the operational manual prior to washing (Table 23). Washing should be done routinely with water or dilute neutral detergents after use. If there is some precipitate that cannot be removed with gentle washing, then the tubes or bottles can be soaked in the dilute neutral detergent followed by washing with a soft brush. Tubes should be rinsed with distilled water and left to air dry. Polycarbonate tubes and bottles are not chemically resistant in alkaline solutions and should not be used with detergents of pH 9 and above. They should not be soaked in any kind of dilute detergents for extended periods as this may cause the material to become brittle.

Sterilization of Centrifuge Tubes and Bottles

In order to sterilize centrifuge tubes and bottles, they should first be washed and dried. Any caps or adaptors should be removed as autoclaving with these on will result in changes in shape of the tubes or bottles.

Storage of Centrifuge Tubes and Bottles

Centrifuge tubes and bottles should be stored at room temperature. After washing, they should beddrained and stored upside down to ensure that all moisture is removed. They should not be exposed to high temperature, humidity, chemical vapours or UV radiation unnecessarily.

Lifespan of Centrifuge Tubes and Bottles

The life span of various tube and bottles depends on a range of factors such as type of samples being separated, speed, temperature, storage and disinfectants used. Tubes or bottles that have exceeded their life expectancy should not be used as continuing to use them could lead to damage of the rotor and centrifuge itself. Tubes and bottles should always be checked for any cracks and deformations prior to use and they should not be used if these deteriorating signs are found.

Table 24: Washing and Sterilization conditions of various centrifuge tubes and bottles

			Tubes and Bottles						Caps, Space caps and crowns			Caps for bottles		Adaptors
Washing	Washing conditions	Parameters	PA	PP	PE	PET	SS	PC	AL	TI	SS	Metal caps	Plastic caps O-rings	
	Washing solution	pH5 or less	n	N	n	n	n	N	n	n	n	n	n	n
	Washing solution	Between pH5 and pH 9	u	U	u	u	u	U	u	u	u	u	u	u
	Washing solution	Above pH9	u	U	u	u	u	N	n	n	n	n	n	n
	Hot water	50 C or less	u	U	u	u	u	U	u	u	u	u	u	u
	Ultrasonic washing	Neutral detergent (pH7)	u	U	u	u	u	U	u	u	u	u	u	u
	Drying conditions	Drying by air												
Sterilization	autoclaving	115 C 30 min	o	O	x	x	o	O	o	o	o	o	o	o
		121 C 20 min	o	O	x	x	o	O	o	o	o	o	o	o
		126 C 15 min	x	X	x	x	x	X	x	x	x	x	x	x
	Boiling	15 – 30 min	o	O	x	x	o	O	o	o	o	o	o	o
		200 – 300 min	x	X	x	x	x	O	o	o	o	o	x	x
	Gas	Ethylene oxide	x	X	x	o	o	O	o	o	o	o	o	o
		Formaldehyde	o	O	o	o	o	X	o	o	o	o	o	o
	Chemical Solutions	Ethanol (70%)	x	X	o	o	o	O	o	o	o	o	o	o
		Hydrogen Peroxide (3%)	o	O	o	o	o	O	o	o	o	o	o	x
		Formalin	o	O	o	o	o	X	x	o	x	o	o	o

Note: PA: polypropylene copolymer, PP: Polypropylene, PE: Polyethylene, PET: Polyethylene terephthalate, SS: Stainless steel, PC: Polycarbonate, AL: Aluminium, TI: Titanium; n = not usable, u = usable, o = satisfactory, x = unsatisfactory (adapted from Hitachi, 2013)

CHOOSING THE APPROPRIATE CENTRIFUGE

A centrifuge is a critical piece of equipment in any laboratory. Since it is such a significant and expensive investment, it is important that buyers make an informed decision to purchase one that best suits their needs and budget. Every laboratory will have different needs regarding space, type of research, frequency of usage or changes in protocols. Ensuring that adequate information is obtained will aid in making informed and long lasting purchases (Goodman, 2007). Regardless of whether you are searching for something flexible that your research center can develop, or a devoted centrifuge for particular protocols, there is a centrifuge out there to meet everyone's necessities. While lab centrifuges have for quite some time been a standard necessity, many individuals still have a great deal of inquiries regarding which centrifuge model is going to be an ideal choice for their necessities. This is not really astounding given the sheer number of lab centrifuge choices accessible in the market today. Picking a centrifuge can be a difficult procedure due to the significant number of models and manufacturers out there.

There are a range of factors that should be taken into consideration before purchasing a centrifuge. Potential buyer should ask the following questions before making a purchase:

- What applications will the centrifuge be applied to?
- What are the maximum and minimum operating conditions?
- What are the volume requirements?
- How many tubes or samples will be processed within one run, one day or one week?
- What kinds of biological samples will the centrifuge need to support?
- What kinds of rotors will be needed to support the applications identified in question 1?
- Will a broad range of protocols be used?

- Will multiple users be using the centrifuge?
- What are the space restrictions? Should I be looking a benchtop or floor space models?
- Will subsequent special needs be required after processing, like biocontainment?
- What is the available budget?

The above questions will provide a better understanding of the buyers needs together with an idea of the kind of centrifuge that can be purchased. Subsequently buyers should also try to gain a better understanding of the basic centrifuge types available. Centrifuges are loosely generally grouped as floor-standing or benchtop models. There are several models and types within each category to choose from depending on the budget, performance requirements and space availability.

Floor Models

Floor model centrifuges are advantageous since they free up bench space. They are often chosen because they are capable of reaching higher speeds or holding higher volumes of samples. Common floor models include ultracentrifuges, and low speed centrifuges.

Bench Top Models

Benchtop centrifuges provide more versatility and offer greater convenience during sample processing. One of the main advantages is that they can accommodate a broad range of user needs thereby making them a more cost effective option. Benchtop options include general-purpose centrifuges, micro-centrifuges, small clinical centrifuges, cell washers, and high-speed models.

ADDITIONAL SELECTION CRITERIA

Selection based on Mechanical Safety

When selecting rotors the following points of mechanical safety should be kept in mind:

- Angle rotors are generally preferred to swing-out rotors because accidents happen frequently with the suspension system of the buckets in swing out rotors.
- Users should be mindful of aerosol production. The tubes of angle rotors should be stoppered properly to minimize aerosol production.
- Rotors should be able to tolerate approximately five times the normal stress without breaking.

Selection based on a Risk of Infection

- The centrifuge and related components should be easy to clean.
- The materials which the centrifuge and rotor is made out of should be able to resist decontaminants, disinfectants and additional mechanical stress.
- Samples in sealed tubes should be placed in sealed buckets to avoid formation of aerosols and leakage when under centrifugation.

Selection based on Electrical Safety

- Compliance with local regulations should be done in order to ensure electrical safety.
- Care should be taken when handling samples
- Carc should also bc takcn aftcr ccntrifugation whcn moisturc condensation occurs to avoid spilling on to switches.

CHAPTER 8

CARE AND MAINTENANCE OF CENTRIFUGES

Overtime the continuous corrosion and contamination of the interior or exterior of the centrifuge machine will affect the lifetime of the equipment. For this reason, strictly following the cleaning and maintenance instructions provided by the manufacturers and suppliers is imperative to ensuring optimal operating conditions and safety of equipment. The balancing of centrifuge tubes is also important as apart from a safety issue, vibration-induced damages to the rotor itself may occur. In addition, the need for replacement of a rotor or centrifuge parts will eventually occur and in order to properly monitor this, it is essential that all users participate in proper usage logs. Accurate recording of run times and centrifugal speeds is important as repetitive acceleration and deceleration of rotors may lead to stress, metal fatigue and ultimately rotor failure.

DAILY MAINTENANCE

Daily cleaning of the inside surfaces of the centrifuge with a suitable neutral disinfectant is important. If a centrifuge tube breaks while in the machine, the area of the centrifuge in contact with the sample should be immediately decontaminated followed by disinfection of the centrifuge bowl. The rotor heads and buckets should be subsequently autoclaved. If any unusual vibrations or noises occur or if any form of deterioration is detected then a qualified service technician should be contacted. Usually strange vibrations or noises are caused by improperly balanced centrifuge tubes which can easily be corrected by rebalancing the samples and tubes.

MONTHLY MAINTENANCE

The outside of the centrifuge, rotor chamber, rotors, and related rotor accessories should be cleaned with a neutral cleaning agent on a monthly basis. Clean plastic and non-metal parts can be cleaned with a new arrangement of 5% sodium hypochlorite (dye) blended 1:10 with water (one section blanch in addition to nine sections water). In addition, the centrifuge speeds should be routinely checked using a tachometer. The accuracy of the centrifuge timer should also be checked and verified according to relevant inspection guidelines and the temperature of refrigerated centrifuges should be checked at least monthly under standardized conditions. Manufacturer's instructions for lubrication, maintenance and replacement of brushes should also be followed. All maintenance steps and checks should be recorded.

ANNUAL MAINTENANCE

The centrifuge should be serviced annually by a qualified service technician who should ensure that the machine is operating safely. This service should include a check and cleaning of condenser coils, fans, screens and any filters. The service technical should also check the centrifuge brushes, bearings, machine timer, temperature, speed and electrical operations. In addition, at the end of the inspection, the technician should issue an inspection certificate indicating compliance with safety laws and proper operation. The most recent service certificate and inspection report should be clearly displayed on the centrifuge. You should be trained and certified in centrifuge operation and upkeep before doing a support process of a centrifuge.

Centrifuge maintenance is basic to protect centrifuge clients. The centrifuges that require standard (month to month) support are Sigma 2-16 (9.129), Eppendorf 5810 (9.136), Sigma 4-16KS (9.134), and Thermo IEC CL10 (8.130a). Additional smaller centrifuges don't require this. All centrifuges get yearly or semiannual expert upkeep. In the event that the centrifuge is especially unclean, report this to an expert as this is unsatisfactory and the client capable must be cautioned.

Basic Method

Continuously wear the proper personal protective equipment when cleaning centrifuges.

1. Operation

Power must be on and the centrifuge top shut to work.

Allow sample to stand for a few minutes.

Balance the centrifuges preceding the start of the run. Adjusting an axis is accomplished by putting tubes with the identical volumes inversely to each other. A tube loaded with water can be utilized if an additional tube is needed for balancing purposes.

Run specimens for 15 minutes at 3200 rpm's to guarantee there is a total gel hindrance between the cells.

2. Cleaning

Clean quickly after any spill to keep corrosives or contaminants from drying on part surfaces.

Month to month cleaning and purifying the inside and outside of the centrifuge is required.

Expel the rotor from the bowl and deliberately clean with a gentle cleanser and flush and dry totally.

Wash the bowl utilizing a mellow cleanser and flush and dry completely.

Expel the containers from the centrifuge and clean with more of the mellow purifying cleanser. In the event that it is essential to utilize 10% Clorox on any metal parts, completely clean with a mellow cleanser and flush to expel the chloride deposit to avoid erosion.

Dry and reinstall basins in the centrifuge.

3. Support

Record all support on the centrifuge Maintenance Log Sheet.

Perform deterrent support and adjustment every year.

Investigate for wear and harm to segments and supplant is fundamental.

Check legitimate adjust, breaking, arrangement and wind stream operation.

Wipe old oil from all rotor sticks and basins with a delicate clean fabric.

Check the centrifuge clock against your look for legitimate operation.

4. Checking rpm

Centrifuge's rpm's ought to be aligned utilizing a tachometer at any rate every year or as required.

Apply around ¼ inch of tape on the middle handle in centrifuge to such an extent that the edge of the intelligent tape divides the focal point of the post.

Close centrifuge top.

Set clock to 6 minutes.

Turn centrifuge on and begin your stopwatch.

Press the power catch on the tachometer to turn on.

Press and hold the power catch while locating light shaft.

Sight the tachometer on the turning intelligent tape.

Discharge the power catch when show has balanced out.

Record the RPM on the new name. Analyze the RPM setting for the manufactory/model of centrifuge.

Record data on the Log sheet.

5. Procedural notes

Rest your hand holding the tachometer over the centrifuge to keep it going.

A reading of 6000 demonstrates a "twofold" perusing. The portion of the tape must be focused, rather than one edge of the tape being focused.

Hold up until the point that the centrifuge stops before applying name.

Set time back to 9-12 minutes

Centrifuges ought to be checked yearly or if there is an issue.

Tachometer battery is a consistent 9 volt battery.

BALANCING YOUR ROTOR

Ensuring that the rotor and respective centrifuge tubes are balanced is imperative for prolonging rotor life and for maintaining good centrifuge operating conditions. Centrifuge tubes should be chosen that are suitable for the desired operating speed and rotor being used. Tubes should be balanced by weighing and should not exceed the limits recommended by the manufacturer. Any hazardous materials should not be placed within 300 mm of the machine (Beckman, 2007). The centre of gravity on opposing containers must be equidistant from the axis of rotation. It should also be of a similar weight and level. The rotor should be correctly positioned on the drive shaft. Some rotors have pins on their underside which aid location and correct alignment in the machine. The buckets on swing out rotors should be assembled correctly and in such a way that they can swing freely. It is also important to remember that a centrifuge user should never walk away from the machine until the required speed has been attained and any unusual vibrations and noises should be investigated immediately.

To balance a rotor, all opposing tube should be filled to the same level with the sample liquid which should have the same density. In addition, the weight of opposing tubes should be distributed equally within the rotor. Tubes should then be placed symmetrically in the rotor (Figure 61). If the amount of sample to be centrifuged is limited and the rotor cannot be balanced, then the opposite rotor cavities can be loaded with centrifuge tubes containing liquids of the same density as the tubes containing the actual samples. In addition, smaller tubes with adapters can be used or thick walled tubes that are partially filled can be used to distribute samples to additional tubes. Low density immiscible liquids such as mineral oil can also be layered on top of the sample to fill opposing tubes so that they are at the same level.

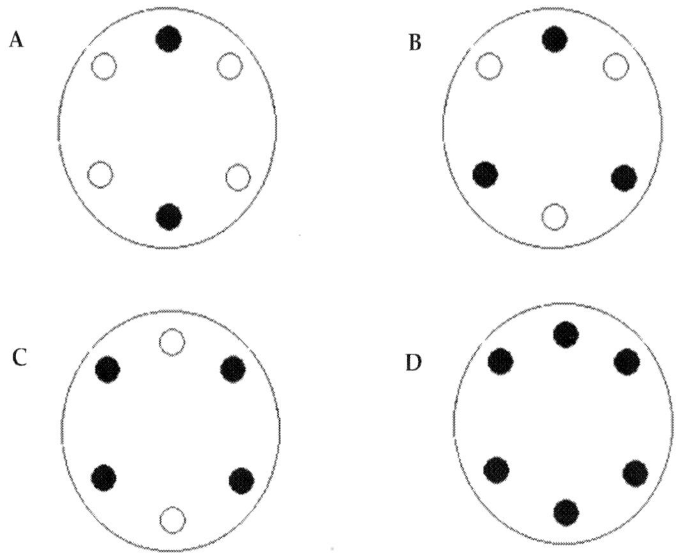

Figure 61: Symmetrical arrangement of centrifuge tubes in a rotor

You don't need to utilize every one of the containers in a rotor. If your samples fit in two containers of the rotor that is totally fine as long as the number and size of samples in each basin are equivalent. You can fill centrifuge tubes with water to enable you to adjust your specimens in the event that you have an odd number of samples to turn.

CHAPTER 9

DESIGNING A CENTRIFUGATION PROTOCOL

When designing a centrifugation protocol a few key factors should be taken into consideration: the denser a biological structure is, the faster it sediments in a centrifugal field and the larger a particle is the faster it moves in a centrifugal field. Next, the denser the buffer or the greater the frictional coefficient, the slower the particle will move. If the centrifugal force is greater, the faster the particles will sediment.

SAMPLE PREPARATION

Before centrifugation can be carried out, depending on the samples to be separated, it may have to undergo a few preparation steps. Firstly tissue samples must be homogenised. The tissue sample is crushed followed by addition of a buffer solution resulting in the formation of a liquid suspension of crushed tissue. Examples of buffer solutions include sucrose and brine. The main purpose of the buffer solution is to serve as a dense and inert solution to suspend the samples without damaging it. This process of breaking down tissues is called homogenization. For density gradient separations, the sample should be ready for loading before the gradient is prepared and should be stored in such a way that it remains cold until the time of the run. The sample is again usually prepared in the same buffer as the gradient. The sample solution itself should have a density that is less than that of the gradient and the run should occur as soon as possible after the sample has been loaded to prevent any 'droplet sedimentation' from occurring. For rate zonal centrifugations, the sample should be loaded on top if the gradient and the sample volume should not exceed 2 to 3 % of the gradient volume. (Yamazaki, 2000).

Filling of Centrifuge Tubes

When filling tubes with samples, an appropriate rack should be used that can securely hold the centrifuge tubes. A pipette or syringe should be used to fill each tube with no fluid being left over in the stem of the tube. Centrifuge tubes should be filled to the appropriately recommended levels as overfilling can lead to an overflow of samples when capping or sealing while too much headspace can lead to deformation of the tube and disruption of and density gradients. In addition, too much air within the tube may require an increased force to remove the tube from the centrifuge cavity after the run.

Capping and Sealing of Centrifuge Tubes

Depending on the centrifuge tubes and protocols being followed, different methods for sealing of tubes will be followed (Table 25). General centrifuge tubes have a uniform cylindrical shape with one end having an opening where the samples are added. Capping and sealing of tubes may be required to prevent any possible leakage of the loaded sample when centrifugation runs are happening. Prevention of leaking is imperative since samples may contain pathogens, mutagens, bacteria or a valuable sample which cannot afford to be lost during spinning. In addition, any leakage will subsequently invalidate the run as a result of rotor imbalances. The capping or sealing mechanism should be fairly tight especially when placed in a fixed-angle rotor as the tubes are usully angled causing an increase in hydrostatic pressure when the centrifuge is rotated at speeds of 20 000 rpm or more. The open end of the centrifuge tube should be inserted properly into the cap in order to achieve a secure seal when tightened. Most often hand tools are used for tightening tubes and capr nghowever, heating methods can also be used. These heat-sealed tubes can only be used once since the seal will have to be cut off in order to remove the sample after centrifugation. Heat sealing is a more expensive method since specialised heaters are required which are often quite expensive.

Table 25: Filling and capping requirements for centrifuge tubes

Sample containers		Filling Levels		
Centrifuge tube materials	**Tube or bottle**	**Swinging bucket rotor**	**Fixed angle rotor**	**Vertical and near vertical tube rotors**
Polyallomer	Thin wall	Filled with 2 – 3 mm headspace	With cap: full	-
	Thick wall	At least ½ full	With cap: full Capless: ½ full to max	-
Polycarbonate	Thick wall tubes	At least ½ full	With cap: full Capless: ½ full to max	-
Stainless steel		Any level	Any level	-
Cellulose propionate		Full	Capless: ½ full to max	-
Polypropylene tubes and bottles		At least ½ full	Capless: ½ full to max With Cap: full	-
Polyethylene tubes		At least ½ full	Capless: ½ full to max With Cap: full	-

Conditions during the Centrifugation Run

Smooth acceleration and deceleration are essential for proper sedimentation of particles especially during gradient centrifugation runs. In addition, accurate temperature control is imperative for reliable and reproducible sedimentation experiments. Most modern centrifuges are equipped with electronic acceleration and deceleration modes which are able to minimize the possibility of disturbances to density gradients (Yamazaki, 2000).

Temperature during the Centrifugation Run

Every centrifuge is manufactured with a specific temperature range that it can withstand. While some models can be run at temperatures as high as 45 °C, many centrifuge tubes and bottles cannot withstand these tenperatures. Most tubes are made of a plastic kind of material that will soften and deform at high temperatures causing poor separations and may even contribute to rotor failure. If a user is not sure if centrifuge tubes can withstand certain temperatures, then they should pre-test the tubes using a buffer instead of the actual sample. Most plastic labware can be centrifuged at temperatures between 2 and 25 °C.

Recovery of samples after Centrifugation

Sample recovery depends on the type of centrifuge and rotor being used as well as the components to be isolated. After a centrifugation run, great care must be taken to avoid disturbances of the pellet or fractionated bands as this could result in breaking of the pellet. In addition, care should be taken to avoid contamination of the fractions to be recovered and this recovery process should be as slow as possible. Usually samples are recovered by decanting the density gradient followed by scraping of the pellet from the bottom. If tubes are to be reused, pellets should be scraped out with a plastic or wooden tool as scratches on the tube interior can result in poor operation of the tubes during subsequent runs. Samples may be retrieved in one of four ways: 1) puncturing the sides of the tube with a needle and syringe just below the banded sample followed by slowly drawing out the sample 2) puncturing the bottom of the tube followed by collection of sample drops 3) removal of the tube plug followed by aspiration of the sample from the open-ended top with a pipette or needle and syringe or 4) slicing of the centrifuge tube using a centrifuge tube slicer.

APPLICATIONS OF CENTRIFUGATION

Centrifugation is one of the most widely applied research techniques utilised in a range of industries such as the mineral, petrochemical, chemical, medical, pharmaceutical, municipal and industrial waste, dairy, food, energy and agricultural industries. They routinely service a variety of disciplines such as biochemistry, molecular biology and medicine, ranging from large-scale commercial applications to laboratory-scale scientific research. Common research applications include the isolation of cells, subcellular organelles, and macromolecules. Industrially, centrifuges are used for a variety of purposes related to the separation of components based on their density. They also service a range of manufacturing industries such as the food, pharmaceutical, biotechnological, environmental and chemical industries. It is also commonly used in sugar and oil refining as well as in the dewatering of vegetables. Many studies have also reported on the advantages of centrifugation for large-scale cell harvesting. Below is a summary of the major applications of the various forms of centrifugation.

Applications of Density Gradient Centrifugation

Separation of Living Cells

For the separation of different cell types making up tissues, firstly the individual cells must be separated from each other by a procedure that is gentle enough to separate cells and still allow them to retain their function. Secondly, the method must allow for different types of cells to be separated from each other. The use of density gradient sedimentation or isopycnic banding can be used for the separation of different cell types while still retaining their function (Table 22). The choice of density gradient medium is important when attempting to separate living cells. Only density gradient solutes of high molecular weights are suitable as they will minimise any instabilities in the sample zone and should be used for the separation of whole cells. Examples of commonly used solutes include bovine serum albumin and Ficoll. For isopycnic bandings, these media may also be used but iodinated media such as Metrizamide would be more suitable as higher viscosity solutions would require longer centrifugation times.

Table 26: Applications of Density Gradients in Isopycnic separations

Gradient Media	Cells	Viruses	Organelles	Nucleoproteins	Macromolecules
Sugars	+	+++	+++	+	-
Polysaccharides	++	++	++	-	-
Colloidal Silica	+++	+	+++	-	-
Iodinated media	++++	++	++++	+++	+
Alkali metal salts	-	++	-	++	++++

++++ Excellent, +++ good, ++ good for some applications, + limited use, - unsatisfactory (adapted from Bisen and Sharma, 2012)

Separation of Cellular Organelles from Mammalian Tissues

Density gradients may also be used to separate different components of a cellular homogenate although this is not recommended as usually not a large amount of material can be loaded onto a density gradient and the separation of entire tissue homogenates is not commonly done. Usually the homogenate is first fractionated by differential pelleting followed by further separation using rate or isopycnic zonal centrifugation. If however a whole homogenate is to be separated then mitochondria, lysosomes and microsomes are separated in the early part of the gradient followed by nuclei and fragments of connective tissue and red blood cells and plasma membranes in the latter part of the gradient.

Applications in Marine Viral Ecology

Viruses are often present homogenously and in large numbers within cell culture lysates. They are smaller than regular bacterial cells or cellular debris from which they need to be separated. Initial methods of separation include filtratio ;ehowever, due to the large amount of cellular debris present in lysed cultures, rapid changes in filter performance may occur thereby reducing the overall amount of viruses collected. As a result of such disadvantages, differential centrifugation using density gradients are often favoured over standard filtration methods. Samples are usually subjected to polyethylene glycol precipitation and ultrafiltration together with subsequent purification steps. Once a relatively pure and concentrated sample is obtained, final purification steps are conducted using density gradients. Purification of viruses from natural samples like marine and freshwater samples is quite similar to protocols of purification from culture lysateserHowever, a few additions steps need to be taken into consideration. The purpose of subsequent downstream processes should be kept in mind when conducting purification steps. For example if the purpose of purification is to allow for subsequent molecular analyses of nucleic acids then more stringent purification methods should be employed to ensure adequate separation from additional cells and dissolved nucleic acids. Common methods again would employ filtration using a membrane with a pore size of 0.2 μm which allows for removal of all cellular debris larger than 0.2 μm. While all viruses will pass through, so will somiamicro bacteria that are smaller than 0.2 μm in diameter

(Bratbak et al. 1992; Garza and Suttle 1995). In certain cases adsorption of some viruses to the filters may also occur thereby reducing the overall efficiency of the method. A more suitable alternative would be the use of buoyant density gradients which very often allow for the simultaneous separation of viruses from cellular and dissolved DNA. In addition since viruses may have the lowest buoyant densities within the sample and the range of buoyant densities of the viruses may overlap with that of existing cells, the separation may not be entirely pure and may be contaminated with other separated layers. To avoid this, an initial filtration step should be employed prior to density gradient centrifugation.

Applications in Stem Cell Research

There are different approaches for the isolation of stem cellr sehowever, these are largely dependent on the tissues in which they are present. One of the more popular techniques for stem cell separation is the use of density gradients to separate cells according to size and buoyant densities. Generally two immiscible solutions with different densities like sucrose and a polymer such as Percoll or Ficoll are combined, forming a system with two layers followed by a heterogeneous cell suspension which is added to the top of the less dense layer. This method is often successfully used for the enrichment of liver stem cells from rate foetal liver cells and for the enrichment of human embryonic stem cells using a monolayer adherent protocol which result in increasing the purity from 30 % to 80 % (Tiago et al., 2013).

Separation of Microalgae

Density gradient centrifugation can be used for the separation of different micro algal strains from mixed cultures. This technique utilises Percoll to achieve separation on the basis of differences in buoyant density. Microalgae form discrete bands at particular positions within the density gradient. Previous studies conducted by Price et al (1972) and Whitelam et al (1983) noted that equilibrium density gradient centrifugation using Percoll can be utilised as a more gentle method for separation than simple pelleting and that each strain has a particular banding density.

APPLICATIONS OF ULTRACENTRIFUGATION

Applications of Analytical Ultracentrifugation

A typical analytical ultracentrifuge can generate a centrifugal field of 250 000 × g. Sedimentation of isolated proteins or nucleic acids through analytical ultracentrifugation is extremely useful for determining the relative molecular mass, purity and shape of these biomolecules. By obtaining such information, detailed insights into the properties of these biomolecules can be obtained. By changing experimental conditions, the effect of variations in pH value, temperature or co-factors on molecular shape can also be studied. Analytical ultracentrifugation is most often employed in determining the purity of macromolecules, determining the relative molecular mass of solutes in their native state, for the examination of changes in the molecular mass of supramolecular complexes, in the detection of conformational changes and in ligand-binding studies. Analytical ultracentrifugation can also be used to estimate sample purity via the sedimentation velocity method.

Determination of Molecular Masses using Analytical Ultracentrifugation

Analytical ultracentrifugation is most commonly used to determine the relative molecular mass of a wide range of biological molecules. For this, small sample sizes of between 20 to 120 mm^3 and low particle concentrations of 0.01 to 1 g/dm^3 are required. Analytical ultracentrifugation can also be used to determine the biochemical properties of biological molecules. Two methods can be used to determine the molecular mass of solutes, namely the boundary sedimentation method and the band sedimentation method. In the boundary sedimentation method, biological particles are initially distributed throughout the solution in an analytical cell. When a centrifugal field is applied, outward migration of the randomly distributed biomolecules will occur throughout the solvent from the centre of rotation. The portion of solvent that no longer has any particles and the portion of solvent still containing particles will form a sharp boundary. The movement of this boundary over time is a measure of the rate of sedimentation of the biological molecules. This sedimentation coefficient also depends directly on the mass of the biological particles and the concertation distribution is dependent on the mass buoyant molecular mass. Alternatively, relative molecular mass of a biological macromolecule can also be determined by band sedimentation in which the sample is layered on top of a denser solvent. During centrifugation, the solvent forms its own density gradient and migration of the band of particles can then be followed in the analytical cell.

Determination of Properties using Sedimentation Coefficients

Conformational and structural changes in biological macromolecules may cause differences in their sedimentation rates which can subsequently be determined by analytical ultracentrifugation. For example, a macromolecule that changes its conformation into a more compact structure decreases its frictional resistance in the solvent. Conversely, frictional resistance will increase when a macromolecular assembly becomes more disorganised. In addition other factors such as the binding of inhibitors or substrates, changes in temperature or buffer conditions may induce further conformational changes in the subunits which in turn result in major structural changes. These major changes can be determined by distinct differences in the sedimentation velocity. Sedimentation equilibrium experiments can also be used to determine the relative size of individual subunits, the stoichiometry and size of a complex of subunits and the strength of interactions between subunits (Wilson and Walker, 2010).

Determination of Molecular Weights

Analytical ultracentrifugation can also be used for the determination of the molecular weight of biological molecules and solutes such as carbohydrates, proteins, nucleic acid macromolecules, viruses and subcellular particles such as mitochondria. This method can be used to determine the molecular weight of molecules ranging in size frol hundredshundred (for example, sucrose) to several million (for example, viruses and organelles). This method is advantageous as it requires small sample sizes and low particle concentrations.

Determination of Sedimentation Coefficients

Biochemical studies over the last few decades have clearly demonstrated that biological macromolecules do not perform their biochemical and physiological functions in isolation. Many proteins have been shown to be multifunctional and their activity is regulated by complex interactions within homogeneous and heterogeneous complexes. Since conformational changes in biological macromolecules may cause differences in their sedimentation rates, analytical ultracentrifugation represents an ideal experimental tool for the determination of such these changes. For example, a macromolecule that changes its conformation into a more compact structure decreases its frictional resistance in the solvent. In contrast, the frictional resistance increases when a molecular assembly becomes more disorganised. The binding of ligands (such as inhibitors, activators or substrates) or a changes in temperature or buffering conditions may induce conformational changes in subunits of biomolecules that in turn can result in major changes in the supramolecular structure of complexes. Such modifications can be determined by distinct differences in the sedimentation velocity of the molecular species. Sedimentation equilibrium experiments can be used to determine the relative size of individual subunits, the stoichiometry and size of a complex assembly under different physiological conditions and the strength of interactions between subunits.

APPLICATIONS IN THE PHARMACEUTICAL INDUSTRY

Production of Biological Products

During manufacturing most biological products remain in colloidal dispersion in water. While it may be difficult to separate these molecules by normal methods of filtration, centrifugation can be used to ease this separation. For example, insulin is purified from other precipitates of protein materials while blood cells are separated from plasma by centrifugal methods. Bacterial enzymes are commonly separate from bacterial culture medium by sedimentation through centrifugation. Dirt and water are also commonly separated from olive oil and fish liver oils via centrifugation methods.

Evaluation of Suspensions and Emulsions

One of the problems when working with suspensions and emulsions is sedimentation and creaming. To enhance the rate of sedimentation and creaming, centrifugation can be carried out between 200 – 3000 rpm. If not sedimentation or creaming occurs by this time, then the suspension or emulsion is regarded as a stable formulation.

Determination of Colloid Molecular Weights

Polymers and proteins often form colloidal dispersions whose molecular weights can be determined by ultracentrifugation. During these runs, larger molecules will collect at periphery while lighter molecules will collect near the centre.

APPLICATIONS IN THE FOOD AND BEVERAGE INDUSTRY

Industrial **centri**fuges are commonly used in the food and beverage industry to separate pulp or pieces of mashed fruit remains from juices. They are commonly used in vegetable and juice purification, wine processing and extractions. Centrifuges are also used to aid clarification of vegetable and animal and dairy products.

APPLICATIONS IN THE ENVIRONMENT

Direct disposal of improperly treated wastewater into the environment can be hazardous to the surroundings if insufficient treatment is applied. Large industrial centrifuges such as decanters are commonly used in water and waste water treatment to remove solids and dry sludge. Decanter centrifuges are commonly used to separate solid materials from liquids and aid in the post-treatment phases of wastewater. Separation of suspended solids occurs and may accumulate on the wall of the bowl which be subsequently removed out of the centrifuge.

CHAPTER 10

TYPICAL SEPARATIONS OF COMMON SAMPLE MATERIALS USING CENTRIFUGATION

SELECTION OF ROTORS FOR CENTRIFUGAL SEPARATIONS

The determination of a rotor efficiency relies upon test volume, number of test particles to be isolated, the molecule measure or potential thickness, wanted run time and wanted nature of partition. The basic rotors sorts include swinging basin, settled edge, vertical and close vertical rotors.

A. Swinging Bucket Rotors

Swinging basin rotors are utilized for pelleting (sedimenting particles from arrangement), rate zonal investigations (division as a capacity of molecule size and thickness), and isopycnic angle considerations (partition as an element of thickness as it were). While all three techniques can be utilized in this rotor, the strategy frequently utilized as a part of swinging pail rotors is the rate zonal thickness slope strategy in which the most extreme determination of test segments is required. Be that as it may, in light of the length of the tubes in the even position, the specimen partition run times are ache for the rate zonal strategies due to the expanded separation the particles are going to go for successful partition.

Also, the isolated zones stay in the same position in the tube in the time of and after centrifugation. The rate zonal strategy isolates particles on the premise of size and also the sedimentation coefficient. In the time of centrifugation, the add up to test weight in tubes is focused on the base (max sweep) of the cans in the swinging-pail rotors.

This weight places abundance weight on the pails in the time of centrifugation and can bring about disappointment of the pails.

B. Settled Angle Rotors (Fixed angle rotors)

Settled point rotors are made with various tube edges from the flat plane, going from 17 to 34 degrees, while the point might be up to 45 degrees in a tabletop or microtube rotor. These edges allow the rotor to securely accomplish the most elevated conceivable speed with the most limited way length or outspread separation for short run times (refer to figure below).

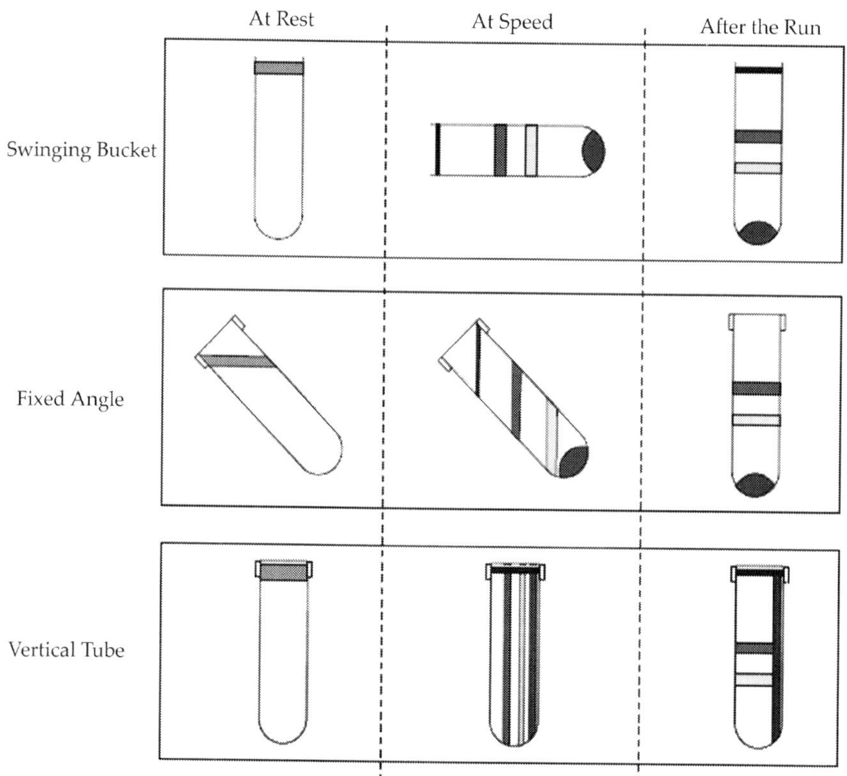

Figure 62: Sample orientation in common rotor types. (Griffith, Applications Note CRI-AN-118, 1996).

These rotors are best utilized for pelleting, yet they are also utilized for density gradient centrifugation. These are the most flexible rotors wherein particles move in a descending way (Coriolis impact) until the underlying residue fills the hemispherical base of the tube (Berman, 1966). The extra sedimenting particles gather on the surface at an edge incidental to the hub of revolution. Studies demonstrate that there is unimportant contrast in the aggregate silt gathered between rotors with various tube points, comparative tube measure, and comparative tube volumes under comparable g-powers (Griffith, Applications Note CRI-AN-118, 1996). An analysis was led that analyzed the most extreme measure of residue that can be gathered in a rotor with an edge of 23 degrees versus the residue gathered in a rotor with a tube edge of 34 degrees. The centralization of the specimen arrangement was sufficiently high to allow the dregs to achieve the tube top. The specimen volumes, tube sizes, and g-powers were comparable for both rotors. The rotor with a tube edge of 23 degrees had add up to silt of 23.3 g/mL, though the rotor with a tube point of 34 degrees had add up to silt of 28.2 g/mL. Next, the grouping of the specimen arrangement in the tubes was lessened to a sum under half of the past run. No distinction in silt volume was seen between the two rotors. Despite the fact that scientists once in a while fill their centrifuge tubes to recoup a full container of silt, this test was done to watch that probability as well. Moreover, the isolated zones change their position in the tube in the time of great many centrifugation. The arrangement volume between zones is more noteworthy to build the partition between the isolated segments.

C. Vertical and Near Vertical Tube Rotors

Vertical tube rotors convey the centrifuge tubes at 0 degrees from the pivot of revolution and close vertical rotors have edges under 9 degrees. Along these lines, these rotors have the most brief way length or outspread separation. In the time of speeding up, the arrangement reorients and sedimentation is done over the breadth of the tube, rather than down the length of the tube as in settled point rotor or on a level plane along the length of the tube as in the swinging container rotor. The shorter way length of the vertical tube rotor decreases the run time for isopycnic angle contemplates. Nonetheless, determination is lost when the detachment includes multi-segment test particles in light of the fact that the silt gathered on the tube divider stays there until the point that the run is ended. After the rotor decelerates to rest, any silt that is pelleted in the time of centrifugation tumbles off the tube divider and retains the isolated specimen zones in arrangement.

SEPARATION THEORY AND METHODS IN BIORESEARCH DISCIPLINES

A. Pelleting and Differential Centrifugation

The most well-known centrifugation technique is differential centrifugation or pelleting. The pelleting method isolates particles of various sedimentation coefficients. Under a consistent diffusive field, the biggest particles in the sample go to the base of the tube first. In the time of pelleting in settled point rotors, the particles will move in a spiral direction down the tube until such point when the underlying dregs reach the base of the tube. This outcome is known as the Coriolis impact (Berman, 1966). Differential centrifugation is best known as the progressive pelleting of particles from a solution, utilizing progressively higher centrifugation powers. In this technique, the centrifuge tube is at first loaded with a uniform blend of test sample. Through every centrifugation procedure, two portions are acquired: a pellet containing the sedimented material and a supernatant arrangement of unsedimented material. A specific part in the blend may wind up in the supernatant or the pellet, or it might be dispersed in both parts, contingent upon its size and additionally the states of centrifugation; see Figure 63.

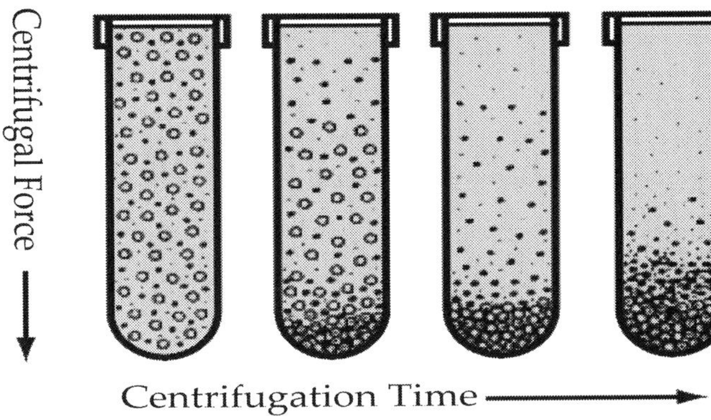

Figure 63: Differential centrifugation (pelleting). (Berman, 1966).

The pellet is a blend of the greater part of the silt segments, and it is tainted with whatever unsedimented particles were at the base of the tube at first. The two divisions from each run are recuperated by emptying the supernatant arrangement from the pellet. This supernatant can be re-centrifuged at higher paces to get another pellet and supernatant. The pellet can be re-centrifuged after resuspension in a smaller volume such as to allow for the "washing" or expelling of the specimen from the pellet. Run times for pelleting can be abbreviated by utilizing half filled thick-walled polycarbonate tubes, conveying smaller tube volumes. A shorter length of tube implies less separation for particles and hence an abbreviated run time. This pelleting technique is generally performed in tabletop and superspeed centrifuges. As this is a preparative system, vast volume rotors have been utilized for such purposes as isolating blood parts and gathering substantial measures of dregs from tissue homogenate or, on the other hand refined bacterial cells from development media. Rotors handling tube volumes of 1.5 mL to 700 mL for each tube with tabletop centrifuges have been utilized routinely for these examinations. With the approach of Fiberlite rotors, sum of the volumes of the specimens from 72 mL to 6000 mL for each can be handled in a story display centrifuge with volumes from roughly 36 mL to 800 mL in ultraspeed centrifuges.

B. Density Gradient Centrifugation

Another common technique for separation is density gradient centrifugation which was first utilized with swinging can rotors for isolating or decontaminating infections, subcellular particles and nucleic acids. Density gradient techniques are especially suited to the investigation of infections. For instance, many studies have shown that infections have been refined and their immaculateness affirmed while sedimentation coefficients and light densities have been measured. In addition, their infectivity has been related with physical properties such as accumulation, separation, and serological contrasts.

Nucleic acids have likewise been considered widely through density gradient techniques. DNA and RNA, and also other poly nucleotides, have been isolated and their sizes assessed from their sedimentation coefficients. Cells and cell particulate divisions, including erythrocytes, spermatozoa, lymphocytes, microbes, spleen cells, mitochondria, and microsomal and other supernatant parts from mammalian tissue, have been disengaged and refined. The density gradient technique includes a supporting section of fluid with density, focus, or consistency that increments towards the base of the tube. This fluid section comprises of an appropriate low atomic weight solute in a dissolvable in which the specimen particles can be suspended.

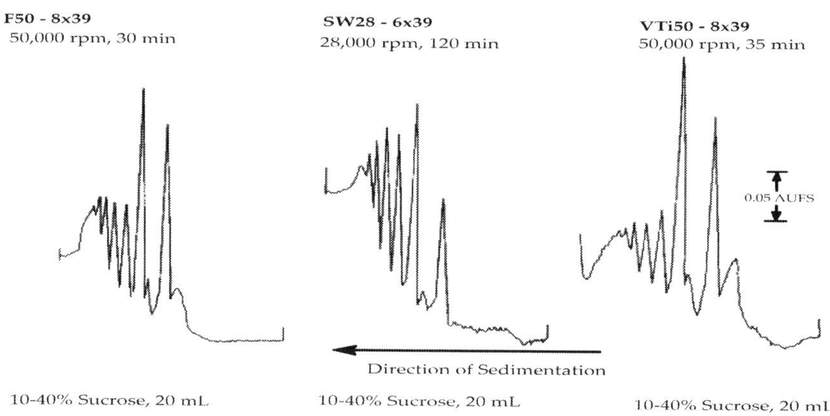

Figure 64: Rate zonal separation of polysomes from rat liver (sucrose gradient). (Brown, 1989).

Average Separations of Common Sample Materials

This segment contains some of the most common partitions of normal tests utilizing the methods already talked about. It ought to be noted that due to the short run length of vertical tube rotors, the greater part of polysomes were not isolated as plainly as in the swinging container and settled point rotors. The sucrose slope fixations were the same in the three rotors and run times were computed from the K-factor of the rotors. M. luteus and λ phage DNA were blended in an answer of 1.65g/mL CsCl and utilized as a part of the three rotors to demonstrate the distinction in detachment. The announced thickness of the λ phage is 1.71 g/mL and 1.73 g/mL for M. luteus. The partition of the DNA's with the settled edge rotor was superior to with the vertical tube and swinging pail rotors (Figure 65).

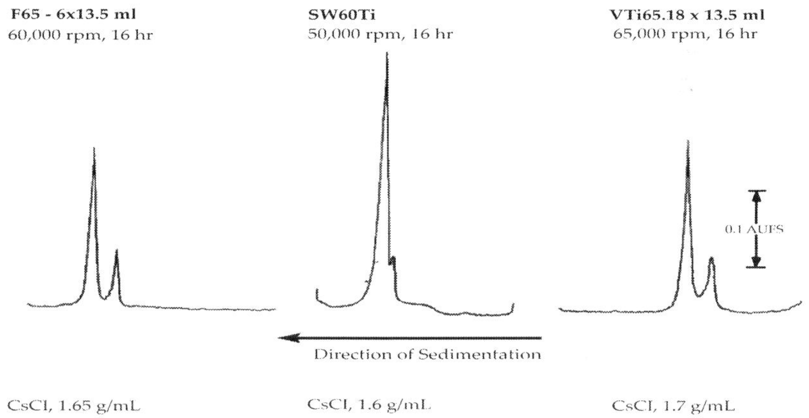

Figure 65: Isopycnic separations of DNA from M. luteus and λ phage. (Brown, 1989).

Isolation of Nucleic Acids

DNA is usually extracted by homogenization of tissues, bacterial cells and viruses using a detergent such as Triton X-100 or sodium dodecyl sulfate (SDS). Additional treatments like boiling or use of alkali chemical solutions can also be used for cellular disruption and isolation of small plasmids. Extraction occurs with phenol or chloroform followed by low speed centrifugation. The DNA is then precipitated using ethanol followed by centrifugation at 13 000 × g (Griffith, 2010). Previous studies conducted by Maniatis et al. (1982) showed the isolation of DNA using a CsCl / EtBr isopycnic gradient method. Ultracentrifugation was done using a force of 200 000 to 300 000 × g for 16 hours. Initial samples were ground in SDS solution followed by the addition of solid CsCl to release the DNA followed by centrifugation. Proteins formed a precipitate at the meniscus while RNA pelleted to the bottom of the tube. Isolated DNA then formed a band in the middle and was collected by puncturing the tube at the zone where the DNA was visible.

Isolation of Bacteria Cells

Bacterial cells grown in tissue culture media for extracting nucleic acids is frequently harvested with conventional super speed centrifuges using fixed angle rotors, which carry bottles with volumes up to 500 ml. The total run time to pellet the bacteria can be up to 20 minutes per run. Standard methods for pelleting begin with inoculation of cultures into a general nutrient rich medium like Luria-Bertani broth (LB) followed by incubation overnight at 37 °C with vigorous shaking. Suggested run times for conventional super speed centrifuges range between 8 – 10 minutes depending on the rotor however, run times of up to 20 minutes can be reached when heavy metal rotors are used (Griffith, 2010).

Standard techniques for pelleting are as per the following: Inoculate Luria-Bertani stock (LB) with *Escherichia coli* containing the plasmid pBR322 and incubate overnight at 37 °C with consistent shaking. Polycarbonate (PC) or polypropylene (PP) containers can be utilized as a part of all huge volume rotors including the Fiberlite F10-6x500y, Fiberlite F9-4x1000y and the Fiberlite F8-6x1000y vast volume rotors. The jugs can be loaded with the microorganisms containing medium and set in the fitting rotors. The proposed set run times for the regular superspeed centrifuges is 8 minutes with the Fiberlite F10-6x500y rotor, furthermore 10 minutes with the Fiberlite F9-4x1000y or F8-6x1000y rotor. Nonetheless, the run times can be up to 20 minutes when overwhelming metal rotors are utilized.

Separation of Whole Cells

Differential gradient centrifugation and density gradients can be used for the isolation of most mammalian cells. The density of most of these cells falls between 1.06 and 1.12 g/ml. previous studies conducted by Leif (1970) and Harwood (1974), showed cells were separated and identified using density gradient centrifugation in floor model centrifuges. Tissue cultures were first subjected to gentle homogenization followed by loading onto density gradients such as colloidal silica, Ficoll, Metrizamide, Nycodenz or OptiPrep. Isopycnic or rate zonal centrifugation is then used to achieve separation although the rate zonal method is usually favoured as in this method cells are subjected to a lower centrifugal force and shorter centrifugation run times.

Isolation of Plasma Membranes

Differential centrifugation is commonly used to isolate subcellular components from tissue homogenates. Many studies report using differential centrifugation to first separate membranes from other components. Tissue samples are first subjected to homogenization followed by centrifugation at low speeds to remove nuclei and other larger debris. This leaves the plasma membranes fragments and mitochondria in the supernatant. Barden et al. (1983) carried out a study in which a technique called density perturbation was carried out. This technique results in changes in the membrane density. The increase in membrane density facilitates the separation of the membranes from other cell components by discontinuous gradient centrifugation. In this method, the nuclei will form a pellet while the plasma membranes and mitochondria will collect at a band together at one of the density steps. Commonly used gradients include sucrose.

Isolation of Lipoproteins

Since different classes of lipoproteins differ in densities and sedimentation, differential flotation is one of the more commonly used methods to separate these classes of lipoproteins. Solutions are centrifuged at approximately 15 000 × g for 30 minutes followed by a subsequent step that allows for the separation of low density lipoproteins which can then be removed at the end of the centrifugation run. A study carried out by Bronzert and Brewer (1977) used an air driven table top ultracentrifuge to quantify cholesterol in plasma lipoprotein fractions.

Conclusion

Centrifugation forms an important technique used when gravity separation, whether by settling, sedimentation or floatation, becomes to slow and particles cannot settle quickly or at all. This technique has come a long way since its commercialization in the 1940s and despite being more than a hundred years old, it still forms one of the most important and widely applied research techniques in biochemistry, cellular and molecular biology and medicine today. Each of these disciplines rely on a particular centrifugation technique for the isolation of cells, subcellular organelles, macromolecules and other components based on size, shape or density, often in high yields. Industrially, centrifugation spans a range of disciplines and industries, from laboratory-scale scientific research to large-scale commercial applications in the mineral, petrochemical, chemical, medical, pharmaceutical, municipal and industrial waste, dairy, food, polymer, energy and agricultural industries. Centrifugation is also used in many manufacturing industries as well, ranging from the food industry where is commonly used for dewatering vegetables to the pharmaceutical and biotechnological industries to aid in the production of bulk drugs and the environmental industries for separation of suspended solids in wastewater. Over the last decade improvements in design has aided advancements in their operations and subsequent applications by allowing for added standard features such as cooling, imbalance detection, noise reduction and electronic programmes. In addition, while centrifuge rotors were initially made of tensile steel, today materials such as aluminium alloys, carbon fibre and titamium are used to allow operation with high centrifugal forces. As research into the properties and structures of biological particles and macromolecules becomes more and more important so does the centrifuge, continuing to be a vital part of any laboratory.

REFERENCES

1. Beveridge T (1997) Juice extraction from apples and other fruits and vegetables. Critical Review of Food Science and Nutrition 37: 449-469.
2. BrunnerKH and Hemfort H (1988) Centrifugal separation in biotechnological processes. In: Mizrahi A (ed.) Downstream Processes: Equipment and Techniques. Advances in Biotechnological Processes, vol 8,pp.1-50. New York: Alan R. Liss.
3. Burgess, N. K., Stanley, A. M. and Fleming, K. G. (2008). Determination of membrane protein molecular weights and association equilibrium constants using sedimentation equilibrium and sedimentation velocity. Methods in Cell Biology, 84, 181–211.
4. Cole, J. L., Lary, J. W., Moody, T. P. and Laue, T. M. (2008). Analytical ultracentrifugation: sedimentation velocity and sedimentation equilibrium. Methods in Cell Biology, 84, 143–179.
5. Cox, B. and Emili, A. (2006). Tissue subcellular fractionation and protein extraction for use in mass-spectrometry-based proteomics. Nature Protocols, 1, 1872–1878.
6. Dahlstrom DA, Bennett RC, Emmett RC et al. (1997) Liquid-solid operations and equipment: centrifuges. In: Perry RH, Green DW and Maloney JO (eds) Perry's Chemical Engineers' Handbook, 7th edn, pp. 18-106. New York, NY: McGraw-Hill.

7. Fisher D, Francis GE and Rickwood D (1998) Cell Separation: A Practical Approach. Oxford: Oxford University Press.
8. Girard, M., Allaire, P. D., Blondeau, F. and McPherson, P. S (2005). Isolation of clathrin-coated vesicles by differential and density gradient centrifugation. Current Protocols in Cell Biology, Chapter 3, Unit 3.13.
9. Graham JM and Rickwood D (1997) Subcellular Fractionation: A Practical Approach. Oxford: IRL Press at OUP.
10. Klassen, R., Fricke, J., Pfeiffer, A. and Meinhardt, F. (2008). A modified DNA isolation protocol for obtaining pure RT-PCR grade RNA. Biotechnology Letters, 30, 1041–1044.
11. LeungWW-F (1998) Industrial Centrifugation Technology. New York: McGraw-Hill. White WF (1979) Centrifuges. In: Bhatia MV and Cheremisinoff PN (eds) Solids Separation and Mixing. Process Equipment Series, vol. 1, p. 81. Westport CT: Technomic.
12. Rickwood, D (1984) Centrifugation: A Practical Approach, 2nd edn. Oxford: IRL Press at OUP.
13. Rickwood, D (1992) Preparative Centrifugation: A Practical Approach. Oxford: IRL Press at OUP.
14. Rickwood, D., && Graham J. M. (2001) Biological Centrifugation. Springer Verlag; ISBN: 0387915761
15. Rickwood, D., Ford, T., & Steensgaard, J. (1994) Centrifugation: Essential Data. 128 pages. John Wiley & Son Ltd. ISBN: 0471942715
16. Whittington PN (1990) Fermentation broth clarification techniques. Applied Biochemistry and Biotechnology 23: 91-121.
17. Anderson, N. G. (1962) The zonal ultracentrifuge. A new instrument for fractionating mixtures of particles. The Journal of Physical Chemistry. 66: 1984-1989.
18. Andritz (2017) Screen bowl decanter centrifuge. Germany. Available at: http://www.andritz.com/se-screen-bowl-decanter-centrifuge-as
19. Barden, A., Lemieux, G., Pallotta, D. (1983) Purification and characterization of plasma membranes from Physarum polycephalum amoebae, Biochimica et Biophysica Acta 730:25-31.
20. Beckman Coulter (2005) Type 70 Ti Rotor - Used In Beckman Coulter Class H, R, and S Preparative Ultracentrifuges.

21. Beckman Coulter. (2007). Rotors and tubes for Beckman Coulter Preparative Ultracentrifuges - User's Manual.
22. Berman, A.S. (1966) Development of zonal centrifuges. National Cancer Institute Mongraph. 21: 41-76.
23. Beveridge, T. (2000) Agriculture and Agri-Food Canada, Pacific Agri-Food Research Centre, Summerland, BC, Canada. Minister of Public Works and Government Services Canada
24. Beveridge, T. (2000). Large-Scale Centrifugation, Minister of Public Works and Government Services Canada. Available at: http://www.biologydiscussion.com/biochemistry/centrifugation/methods-used-for-separation-of-particles-in-centrifugation-3-methods/12453
25. Birnie, G. D. (1972). Separation of native and denatured DNA, RNA and hybrid on sodium iodide gradients. 27: 19-22. Available at: https://doi.org/10.1016/0014-5793(72)80399-5
26. Borchard, W. and Staatmann, A. 2003. Analytical Ultracentrifugation VI. Science. Available at:
27. https://books.google.co.za/books?id=Yv5pCQAAQBAJ&dq=preparative+ultracentrifugation&source=gbs_navlinks_s
28. Brakke, M.K. (1958) Estimation of sedimentation constants of viruses by density-gradient centrifugation. Virology. 6: 96-114
29. Brakke, M. K. (1960) Density gradient centrifugation and its application to plant viruses. Advances in Virus Research. 7:193-224.
30. Bratbak, G., Haslund, O. H. Heldal, M. Noess, A and Roeggen T. (1992). Giant marine viruses. Marine Ecology Progress Series. 85:201-202.
31. Bronzert, T. J., Brewer Jr., H. B. (1977) New micro-method for measuring cholesterol in plasma lipoprotein fractions. Clinical Chemistry. 23:2089.
32. Catalano, S., Palazzolo, J., Robertson, M., Wesorick, S., Nalbandian, A., & Minbiole, K. (2017). Encyclopedia of Chemical Engineering Equipment – centrifuges. College of engineering. University of Michigan. Available at: http://encyclopedia.che.engin.umich.edu/Pages/SeparationsMechanical/Centrifuges/Centrifuges.html
33. Centrifuge World (2017) How a pusher centrifuge works. Available at: http://centrifugerepairshop.com/pusher-centrifuge-works/

34. Claude, A. (1946). Fractionation of mammalian liver cells by differential centrifugation. Problems, methods and preparation of extract. The Journal of Experimental Medicine. 84(1): 51–59.
35. Creighton, Thomas E.. (2010). Physical and Chemical Basis of Molecular Biology - 16. Sedimentation by Centrifugation. Helvetian Press.
36. Available at: http://app.knovel.com/hotlink/pdf/id:kt0091X9N1/physical-chemical-basis/sedimentation-bycentrifugation
37. Dimauro, I., Pearson, T., Caporossi, D., & Jackson, M. J. (2012) A simple protocol for the subcellular fractionation of skeletal muscle cells and tissue. BMC Research Notes. 5:513. Available at: http://www.biomedcentral.com/1756-0500/5/513
38. Fernandes, T. G, Diogo M. M, Cabral, J. M. S. (2013) Stem cell bioprocessing: For cellular therapy, diagnostics and drug development. Elsevier Science: 236. ISBN 10: 1907568883 ISBN 13: 9781907568886
39. Garza, D. R., and Suttle, C. A. (1995). Large double-stranded DNA viruses which cause the lysis of a marine heterotrophic nanoflagellate (Bodo sp.) occur in natural marine viral communities. Aquatic Microbial Ecology. 9:203-210.
40. Graham, J., & Rickwood, D. (year) Biological Centrifugation: Centrifugation, A Practical Approach (2nd ed.), p.12
41. Griffith, O. M. (1996) Applications Note CRI-AN-118, Mountain View, CA, Composite Rotor.
42. Griffith, O. M. (2009) Sample Re-mixing during Density Gradient Separations with Thermo Scientific Fiberlite F21-8x50y mL Fixed-Angle Rotor. Technical Note, AN-LECF-CFSAMPLEREMIX-0908.
43. Griffith, O. M. (2010). Practical Techniques for Centrifugal Separations. Applications guide. Thermo Fisher Scientific Inc.
44. Harwood, R. (1974) Cell separation by gradient centrifugation, International Review of Cytology. 38: 369-403.
45. Hitachi, K. (2013). Instruction Manuel. Rotors, Tubes, Bottles and Caps. For Hitachi Preparative Ultracentrifuge.
46. Holtbrügge, J., Kunze, A. K., Niesbach, A., Schmidt, P., Schulz, R., Sudhoff, D., & Skiborowski, M. (2016). Reactive and Membrane-

Assisted Separations. P. Lutze, & A. Górak (Editors). Walter de Gruyter GmbH & Co KG.

47. Lawson, V. A., Stewart, J. D., & Masters, C. L. (2007). Enzymatic detergent treatment protocol that reduces protease-resistant prion protein load and infectivity from surgical-steel monofilaments contaminated with a human-derived prion strain. Journal of General Virology. 88: 2905–2914.
48. Lawrence, J. E. & Steward, G.F. (2010). Purification of viruses by centrifugation. Aquatic Viral Ecology. Ed.. S W. Wilhelm, M G Weinbauer and C A Suttle. pp166 – 181. Waco, TX: ASLO.
49. Leif, R.C. (1970) Buoyant density separation of cells, Automated Cell Identification and Cell Sorting, 21-96, ed. G.L. Weid and G.F: Bahr, New York, Academic Press.
50. Leung, W. W. F. (2007). Centrifugal Separations in Biotechnology. Technology & Engineering. Academic Press. 312 pages. Available at:
51. https://books.google.co.za/books?id=Oci4GTqikk4C&dq=%22batch+feed+centrifuges%22&source=gbs_navlinks_s
52. Lindahl, P. E. (1948). Principle of a counter streaming centrifuge for the separation of particles of different sizes. Nature (London) 161 :648-649.
53. Majekodunmi, S. O. (2015) A Review on Centrifugation in the Pharmaceutical Industry. American Journal of Biomedical Engineering. 5: 67-78. DOI: 10.5923/j.ajbe.20150502.03
54. Maniatis, T., Fritsch, E.F:, Sambrook, J. (1982) Molecular Cloning: A Laboratory Manual, New York, Cold Spring Harbor Laboratory.
55. McEwen, C. R., Stallard, R. W., and Juhos, E. T. (1968). Separation of biological particles by centrifugal elutriation. Analytical Biochemistry. 23:369
56. Muze Development (2016) How a decanter centrifuge works. Vision Machine Inc. Available at: http://www.visionmachine.com/how-a-decanter-centrifuge-works/
57. Open Learn University. 2017. Royal Charter (RC 000391) Available at: http://www.open.edu/openlearn/science-maths-technology/science/tour-the-cell/content-section-2.2

58. Parish, J.H. (1972) Principles and Practice of Experiments with Nucleic Acids, 131-155, London, Longman.
59. Pertoft, H., Philipson, L, Oxelfelt, P and Höglund, S. 1967. Gradient centrifugation of viruses in colloidal silica. Virology 33:185-196.
60. Price, C. A. (1972) Zonal centrifugation. Manometric and Biochemical Techniques. (W.W. Umbreit, R. H. Burris, and J. F. Stauffer, eds.) 5th ed. Vol 11, pp. 213 – 243. Burgess, Minneapolis, Minnesota.
61. Ralstone, G. Introduction to analytical ultracentrifugation. Beckman. Department of biochemistry, the university of Sydney. Sydney, Australia.
62. Rickwood, D. (1984). Centrifugation, a practical approach: The theory and practice of centrifugation. p1 – 43. IRL Press, Oxford, England.
63. Rickwood, D. (1992) Centrifugation: A Practical Approach. ASIN: 090414755X.
64. Rickwood, D. (2014). Centrifugation Techniques. Available at: https://www.researchgate.net/publication/227982944
65. Rickwood, D., Ford, T., and Steensgaard, J. (1994) John Wiley & Son Ltd. ISBN: 0471942715
66. Schroeder, S. 2013. Routine Maintenance of Centrifuges Cleaning, Maintenance and Disinfection of Centrifuges, Rotors and Adapters. Eppendorf AG, Hamburg, Germany.
67. Schumaker, V. N. (1967) Zone Centrifugation. Advances in Biological and Medical Physics. 11:245-339.
68. Sigma-Aldrich Co. (2011). Biofiles - Centrifugation. Volume Number: 6, No 5.
69. Sykes, J. (1971) Centrifugation techniques for the isolation and characterization of subcellular components from bacteria. Methods in Microbiology, 5:55-207.
70. Records, A., & Sutherland, K. (2001) Decanter Centrifuge Handbook. Science. ISBN 1 85617 3690. Oxford. Elsevier Advanced Technology.
71. Sutherland, K. (2005) Centrifuge focus: solids removal - the options. Filtration and separation. 42:16-20.

72. Steward, G. F., Montiel, J. L. & Azam, F. (2000). Genome size distributions indicate variability and similarities among marine viral assemblages from diverse environments. Limnology and Oceanography. 45:1697-1706
73. Svedberg, and Pedersen, K. O. (1940). The Ultracentrifuge. J. Phys. Chem. 44:(7), pp 952–953
74. DOI: 10.1021/j150403a021
75. The Biomedical Scientist. 2013. Centrifugation: a guide to equipment use and maintenance. Back to basics. Available at:
76. file:///C:/Users/Shalinee/Desktop/Centrifugation%20Techniques/centrifugation.pdf
77. Vinograd, J. (1963) Methods in Enzymology. 6:854-870, ed. S.P. Colowick and N.O. Kaplan, New York, Academic Press.
78. Voëlkl, A. (2010). Ultracentrifugation. Encyclopedia of Life Sciences. DOI: 10.1002/9780470015902.a0002969.pub2
79. Vogel, HC., & Todaro CM. (1996) Fermentation and Biochemical Engineering Handbook, Principles, Process Design and Equipment. (2nd Ed) William Andrew. Technology & Engineering - 828 pages
80. Wakeman,R.J.(2011)Centrifuges.DOI:10.1615/AtoZ.c.centrifuges. Available at: http://www.thermopedia.com/content/620/
81. Whitelam, G.C., Lanaras, T. & Codd, G.A. (1983) Rapid separation of microalgae by density gradient centrifugation in Percoll. British Phycological Journal. 18:23-28. DOI: 10.1080/00071618300650031
82. Whittington, P. N. (1990) Fermentation broth clarification techniques. Applied Biochemistry and Biotechnology 23: 91-121.
83. Wilson, K and Walker, J. (2010) Principles and Techniques of Biochemistry and Molecular Biology (seventh edition)
84. Wilson, Ian D. Poole, Colin F.. (2009). Handbook of Methods and Instrumentation in Separation Science. Volume 1 - Centrifugation to Covalent Chromatography. Elsevier. Available at: http://app.knovel.com/hotlink/pdf/id:kt00BYXIS1/handbookmethods-instrumentation/centrifugation-covalent
85. Wilson K., and Walker, J. (2009) Principles and techniques of biochemistry and molecular biology. 7th ed.

86. Xu W., DeCroix, D. S. & Sun, X. (2014). Mechanistic based DEM simulation of particle attrition in a jet cup. 253:385–392. Available from: https://www.researchgate.net/publication/275071937_Mechanistic_based_DEM_simulation_of_particle_attrition_in_a_jet_cup
87. Yamazaki, A., (2000) Kresge Eye Institute, Wayne State University, Detroit, MI, USA Copyright 2000 Academic Press
88. Allen, T. (1968). *Particle Size Measurement.* London: Chapman and Hall.
89. Backhed, F., Ley, R., Sonnenburg, J., Peterson, D. A., & Gordon, J. (2005). *Host-bacterial mutualism in the human intestine.*
90. Berman, A. (1966). Nat. Cancer Inst. Mongr.
91. Brestoff, J. R., & D.Artis. (2013). *Commensal bacteria at the interface of host metabolism and the immune system.*
92. Brown, R. (1989). *The physics of continuous flow centrifugal cell separation.*
93. Buffie, C. (2015). Precision microbiome reconstitution restores bile acid mediated resistance to Clostridium difficile.
94. Caporaso, J. G. (2010). QIIME allows analysis of high-throughput community sequencing data.
95. Eiseman, B., Silen, W., Bascom, G., & Kauvar, A. (1958). *ecal enema as an adjunct in the treatment of pseudomembranous enterocolitis.*
96. Faith, J. (2013). *The Long-Term Stability of the Human Gut Microbiota.*
97. Ford, T. C., & Graham, J. M. (1991). *An Introduction to Centrifugation.* BIOS Scientific Publishers.
98. Gill, S. (2006). *Metagenomic analysis of the human distal gut microbiome.*
99. Griffith, O. (1996). *Applications Note CRI-AN-118.* CA.
100. Griffith, O. (2007). *Density Gradient Separations in Vertical Tube, Near Vertical (NVT), Fiberlite Fixed Angle and Swinging Bucket Rotors: A Comparative Study.* Piramoon Technologies.
101. Haas, B. J. (2011). Chimeric 16S rRNA sequence formation and detection in Sanger and 454-pyrosequenced PCR amplicons. Genome Research .

102. Hamilton, M. J., Weingarden, A., Sadowsky, M. J., & Khoruts, A. (2012). Standardized frozen preparation for transplantation of fecal microbiota for recurrent Clostridium difficile infection. *American Journal of Gastroenterology*, 761–767.

103. IDF. (1982). Determination of the total solids content. Reference method. Standard FIL-IDF 4A:1982. International Dairy Federation. Brussels, Belgium.

104. Jacobsen, C., & Rasmussen, O. (1992). *Development and application of a new method to extract bacterial-DNA from soil based on separation of bacteria from soil with cation-exchange resin.* Applied and Environmental Microbiology.

105. Jones, T. P., & Rowe, N. P. (1999). *Fossil Plants and Spores: Modern Techniques.*

106. Kengl, P., FigdorJack, LeemansWilly, M., BontJan, S., & Vries, E. d. (1983). *Theory and practice of centrifugal elutriation.*

107. Lindahl, V., & Bakken, L. (1995). Evaluation of Methods for Extraction of Bacteria from Soil. Fems Microbiology Ecology.

108. Makhzami, S. (2008). In situ gene expression in cheese matrices: Application to a set of enterococcal genes. *Journal of Microbiological Methods*, 485–490.

109. McDonald, D. C., Kuczynski, J., Rideout, J. R., Stombaugh, J., Wendel, D., Wilke, A., et al. (2012). *he Biological Observation Matrix (BIOM) format or: how I learned to stop worrying and love the ome-ome.*

110. McEwen, C. R., Stallard, R., & Juhors. (1968). *Anal Biochem.*

111. Milani, C. (2013). *Assessing the Fecal Microbiota: An Optimized Ion Torrent 16S rRNA Gene-Based Analysis Protocol.*

112. Morgan, X. (2012). *Dysfunction of the intestinal microbiome in inflammatory bowel disease and treatment.*

113. Murayama, K., Fujimura, T., Morita, M., & Shindo, N. (2001). One-step subcellular fractionation of rat liver tissue using a Nycodenz density gradient prepared by freezing-thawing and two-dimensional sodium dodecyl sulfate electrophoresis profiles of the main fraction of organelles.

114. Persidsky, M., & Olson, L. (1978). *Biology Medicine.*

115. R. Grabske. (1978). *News of Biochemical Instrumentation, Spinco.*

116. Regnéll, J., & Everitt, E. (1996). *Preparative centrifugation — a new method for preparing pollen concentrates suitable for radiocarbon dating by AMS.*

117. Rickwood, D. (1992). *Centrifugation: A Practical Approach.*

118. Rickwood, D., & Graham, J. M. (2001). *Biological Centrifugation.* Springer Verlag.

119. Rickwood, D., Ford, T., & Steensgaard, J. (1994). *John Wiley & Son Ltd.*

120. Rooijers, K. (2011). *An iterative workflow for mining the human intestinal metaproteome.* BMC Genomics .

121. *Service and Support | Beckman Coulter*. (2000). Retrieved from Beckman: https://www.beckman.com/support

122. Torsvik, V. (1980). *Isolation of bacterial DNA from soil.* Soil Biology & Biochemistry .

123. Weast, R. C. (1974-1975). *CRC Handbook of Chemistry and Physics: A Ready-reference Book of Chemical and Physical Data.* CRC Press.

INDEX

T

V

Z